GIANT
PANDA!

你不知道的大熊猫
The Giant Panda You may not Know

大熊猫身份证
Giant Panda's ID Card

姓　名

大熊猫

别　名

食铁兽、花熊、黑白熊

学　名

Ailuropoda melanoleuca

英文名

Giant Panda

出生地

中国

07

大熊猫已在地球上生存了至少
800 万年，被誉为 "活化石" 和
"中国国宝"。

大熊猫属于
食肉目
熊科
大熊猫亚科
大熊猫属
是该属唯一的哺乳动物。

中国国家重点保护野生动物
名录：国家 I 级保护野生动物。

"濒危野生动植物种国际贸易公约"
(CITES) 附录 I 物种。

世界自然基金会的会徽、
会旗中的标志，是全球生物多样性
保护的旗舰物种。

The Origin
of Species

起源

GIANT
PANDA!

地球的演变历史纷繁复杂，先后存在过数亿物种，许多物种在漫长岁月中被自然法则淘汰，也有一些物种通过进化，逐渐适应大自然而绵延至今，大熊猫就是其中之一。从始熊猫开始，大熊猫的体型演变是由小到大，再变小，历经 800 万年。

6500 万年前，恐龙灭绝

约 800 万年前，大熊猫出现 始熊猫属

约 200 万年前，小种大熊猫

约 100 万年前，巴氏大熊猫（体型最大）

约 1.2 万年前，现生大熊猫

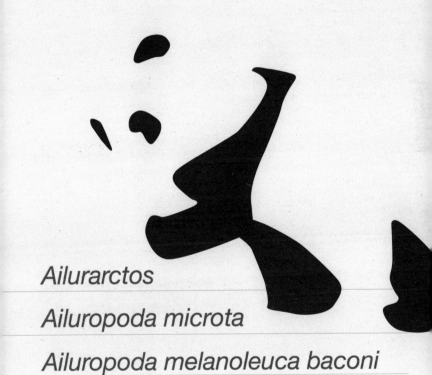

Ailurarctos

Ailuropoda microta

Ailuropoda melanoleuca baconi

Ailuropoda melanoleuca

6500 万年前
恐龙灭绝

始熊猫
（约 800 万年前）

小种大熊猫
（约 200 万年前）

巴氏大熊猫
（约 100 万年前）

现生大熊猫
（约 1.2 万年前至今）

The Legend
of Giant Panda

大熊猫的故事

GIANT
PANDA!

01
生命森林
Forest of Life

物种进化就像是一张蜘蛛网，一个物种灭绝了，就会出现一个空洞的点，而以这个点为起点，又会再次引发其他点的消失……你眼睛看得到的是一个物种的灭绝；你看不到的，可能就是一个生态系统的崩塌。

人类能与大熊猫生活在同一个世界，演化历程发生交错，是我们的运气。在世界众多物种中，大熊猫生存在地球上超过 800 万年，保留了漫长的演化过程和独特的基因密码。

在横断山的东端，有一条介于长江和黄河的狭长的绿色走廊，就是大熊猫走廊带。大熊猫走廊带纵横中国的四川、陕西和甘肃三省，这条走廊也是世界上唯一的大熊猫栖息地。

这里是中国乃至全世界生物多样性最丰富、最集中的地区之一，丰富多样的生物环境使得各种生物都能在横断山区找到属于自己的家园。在这里，崇山峻岭中，山高、涧深、人迹难至，也使得大熊猫这一古老物种最终得以以一种隐蔽的形式存活下来。

这里海拔约 2000~3700 米的森林中不仅雨量充沛，气候湿润，竹林遍布，终年云雾缭绕，温度常年在 20℃以下。而竹子是众多的植物中唯一一种可以在一年四季里为大熊猫提供能量的食物。

02
小小奇迹
Little Miracle

一只成年大熊猫的体重为90~130千克，而一只初生的大熊猫幼仔平均体重约为100克，是妈妈体重的千分之一。很难想象这样一个"超级小不点"可以渐渐地长成"超级大个子"，它的生存完全依赖大熊猫妈妈的细心照护。

紧闭着眼睛，没有听觉也没有视觉，只有微弱的嗅觉，这就是一只大熊猫幼仔刚刚出生的样子。刚刚出生的大熊猫幼仔不是黑白相间，而是浑身粉红，只有一层薄薄的绒毛。

7 天后，它的眼圈、耳、前肢、肩部皮肤的位置才开始微微变黑，白色绒毛开始变密。

40 多天，大熊猫幼仔才逐渐睁开双眼，开始认识这个世界。

60 多天，开始有听觉。

90 多天，大熊猫幼仔已经和我们见到的大熊猫一模一样了，它们开始学走路。

120 多天后的大熊猫幼仔开始调皮地翻滚、嬉闹、爬树，学吃竹子。

到了 180 天，它们的体重开始迅速增加，随着身体的长大，它们的活动范围也越来越大。

与自然界的每个生命一样，它们都是大自然的奇迹。春天，刚刚受孕的大熊猫妈妈开始四处觅食，通过不断地进食，来满足身体所需的营养。在怀孕的 5 个月里，它们会在初春时吃竹笋，之后慢慢开始吃竹叶。这个时期的大熊猫妈妈进食量非常大，都是为了幼仔在肚子里健康成长。在此期间，它们还会在海拔高一点的地方寻找洞穴。洞穴的位置一般在大树底部的空心树洞里，或者巨大的岩石下，周围必须要有充足的水源和竹林。一旦确定好位置，它们就会寻找许多的竹节还有枝叶铺在洞底，用竹叶和苔藓筑成柔软的巢穴，以确保幼仔的舒适。

在幼仔出生的 14 天里，大熊猫妈妈基本处于禁食状态。幼仔满月前，大熊猫妈妈要一直把它抱在怀里。无论它们感觉饿了、渴了，还是冷了、热了，或者妈妈把它们抱得稍稍紧了，它们都会主动地发出不同的声音，提醒妈妈要好好地、更加精心地照顾它们。

03
幼时记趣
Scenes from Childhood

对于大熊猫幼仔来说，童年时光是最重要的也是最快乐的，它们在跟妈妈度过的这一年里，不但要学会许多生存技巧，还要去认识周遭的生活环境。它们将走过森林的许多地方，有时在山坡上奔跑，有时在溪流中嬉戏。

大熊猫幼仔出生后的第三个月，妈妈就开始带着它在家门口活动了。刚开始，大熊猫幼仔还不会走路，妈妈会用嘴衔着它四处行走。学会自己行走的大熊猫幼仔对周边的环境充满好奇。这个时候，妈妈就开始教它第一课，练习爬树和觅食。有时它会在山坡上逗留，多玩一会儿，然后在路上凭借气味辨别妈妈的方向。大熊猫妈妈有时候也会在不远处观察。如果迷路或者气味标识被破坏了，幼仔就会发出尖脆的叫声，等待妈妈的引领。

大熊猫妈妈会告诉幼仔，树枝是避险的好地方，如果遭遇危险，可以爬上树木，用自己的前爪牢牢抱住树干，后爪则用力攀爬。大熊猫妈妈也会带着幼仔，四处游历，它要知道自己生活在什么地方，附近哪里有可口的食物和清洁的水源。

大熊猫妈妈还会教幼仔辨别哪些竹子好吃，哪些竹子口味生涩。

大熊猫通常能够在不同的季节辨别出哪种竹子最好吃、竹子的哪一部分最有营养、最合口味。

夏天，大熊猫妈妈会带着幼仔去高海拔地区乘凉避暑；冬天，它们也不会冬眠，厚厚的绒毛帮助它们御寒，它们穿行在风雪中，寻找食物。

幼年时的大熊猫非常顽皮活泼，在野外会坐着滑坡或滑雪，个别少年大熊猫到冬天还会下山串户走村，把村民们的水桶器皿搬出村舍当玩具，尽情玩耍之后又丢在山野之中。在它们眼中，什么都可以当作玩具。吃饱饭之后，它们除了睡觉就是玩。

随着时间的推移，大熊猫幼仔的各个器官机能逐渐完善，眼睛周围的黑色眼圈已经长到了 5 厘米，像是戴着一副墨镜，一对黑色的耳朵又大又圆。它的眼睛很小，视力比听力要差一点。黑色的耳朵是对寒冷环境的适应，能吸收热能，加快末端血液循环，并减少热量的散失。大熊猫是怕热大过于怕冷，它们喜欢生活在凉爽的地方。

大熊猫幼仔在这一年中，食量渐渐变大，每天觅食的时间也逐渐变长，而这一年短暂而充实的锻炼，让大熊猫幼仔快速成长，体格变得强壮，体型也日渐庞大，到了去闯荡世界的时候了。

04
何以为家
Home of My Own

对于所有物种来说，成熟就意味着要离开妈妈，开始独立生活。一种内力在它的生命中产生，那便是离开母亲的倾向，有如候鸟，季节一到，就一定要启程远飞。

但是，大熊猫对从小生活的家园，往往难舍难分。为了避免近亲繁殖，保证繁衍的延续性，大熊猫妈妈不得不将女儿驱赶出去，把儿子留在附近。

每只大熊猫都会有 4~7 平方千米大小的家园，这是它们在考察了不少地方之后选择的区域。家园附近需要有繁茂的竹林，丰盛的食物，还有溪流和清泉，并且也很安静和隐蔽，适合它们生活。虽是一方小小天地，却可以安心休憩，谈情说爱，养儿育女。建立家园对于大熊猫来说是非常重要的。

广阔的生活环境让大熊猫寻找一块合适的区域似乎没有那么难，它会在家园的边界处留下气味，这样可以避免与其他大熊猫发生冲突。

大熊猫在两岁半的时候，会感觉到妈妈对它的疏远，有时还会伴有驱赶的行为。妈妈的这种行为，意味着它该离开这片区域，寻找和建立自己的家园了。即便是被迫离开了这个区域，它也还是会不断地回到这里来；即便建立自己的新家，也不会离这个范围太远。不管到了哪里，它都可以凭借气味找到曾经的家园。这个离别的过程虽然残酷，却也是大熊猫能够长久生存下来的原因之一。

以固定的领域作为家园的大熊猫，是在长时间的演化当中，找到的一种生存智慧。每只大熊猫各守一方土地，既可以调节种群的密度，也可以避免大熊猫之间为了食物而产生的竞争。这些更深远的智慧，需要大熊猫幼仔在建立家园和拥有自己的家人之后，才能逐渐领悟和学习到。

05
竹海游侠
Rangers in the Bamboo Forest

世界万物都值得我们欣赏与尊重，但是大熊猫的意义，似乎更有其独特的价值。它的外形憨厚可爱，惹人喜欢；它的性格与世无争，引人侧目；而它在漫长的物种进化中，从"食肉"渐渐地进化转变为吃竹子，避免了与人或其他动物争夺食物造成的伤害，充分体现了这一古老生物的高超智慧。但是这些对于大熊猫来说，并不重要，在野外自由自在地生活，才是它们最喜欢的事情。

食不分昼夜，睡不择栖处，排泄也无固定地方。它们爱游荡，没有固定的觅食处和休息地，在竹海里游荡觅食，喜欢自由自在的生活。在绝大部分时间里，大熊猫都生活在高山密林之中，独来独往，难觅其踪。

一只成年的大熊猫体重可达 90~130 千克，站起来的身高约 1.7 米，看起来身材庞大，行动迟缓，其实大熊猫的行动十分灵活，无论是觅食还是爬树，都是高手，只要需要，大熊猫在林下能够以每小时 40 千米以上的速度奔跑。

大树，对于大熊猫来说，不仅仅是躲避危险的地方，还是休息放松、玩乐的地方。对于从小深谙爬树之道的大熊猫来说，它们全身的关节都非常灵活，可以像柔术演员一样完成高难度动作。在爬树的时候，前肢贴着树身，慢慢向上，直到后肢直立，然后爬上大树。在下树时头朝上，慢慢下退，直到接近地面时，才敏捷地一跃而下。

它们在树上，可以呼吸更加清新的空气，享受和煦的阳光，还可以在树上居高临下，下面有什么风吹草动，放眼一看，便了如指掌。有时候它们整天甚至几天待在树上，或睡或坐。当遇到天敌或者相互追逐嬉戏的时候，它们也可以很快地爬到高高的树上去，对手对此也毫无办法。

对它们来说，在树上的生活，并没有因为体型的问题受到太多的限制。它们可以游刃有余地在高大的、令人目眩的树梢上，荡来荡去，却毫不畏惧，每当以为它们快要掉下来的时候，它们总能一次又一次地化险为夷。在树上睡觉的时候，它们也会找到一个舒适的树干分叉处，骑着坐，抱着睡，甚是惬意和享受。

天生好水

中国有句古语：智者乐水，仁者乐山。在山林中生活的大熊猫，就像是一位隐退的智者。密林中的小溪边，或者潺潺流动的泉水边，时常可以看到大熊猫的身影。

一只大熊猫要建立生活区，就得在 800 米以内找到一处水源。为了找到干净可口的饮用水，大熊猫常常不惜长途跋涉，从高高的山上来到河谷之中，一旦到达水源处，便开怀畅饮。有时不顾姿态躺卧在水边，拉开架势，一口气喝足之后，就开始与同伴嬉戏打闹，玩得不亦乐乎。夏天在水中洗澡，冬天也可以在冰天雪地里冬泳。

大熊猫可以游山玩水，可以横渡湍急的河流，也可以在遇到危险的时候，向远处疾跑；它们既笨重又灵活，既憨厚又机敏，是不可貌相的动物。

素食之路

生活在野外的大熊猫，总是能找到一条充满智慧和富有生趣的生存之路。几百万年前，为了摆脱与其他食肉动物的竞争，它们选择了一条素食之路，开始以竹子为主要食谱。而选择这样的饮食习惯，需要付出的代价是，它们得把一生的大部分精力放在吃饭上。

大熊猫一天的大致作息时间

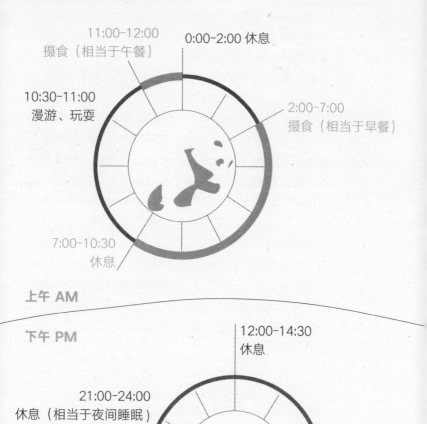

11:00-12:00
摄食（相当于午餐）

0:00-2:00 休息

10:30-11:00
漫游、玩耍

2:00-7:00
摄食（相当于早餐）

7:00-10:30
休息

上午 AM

下午 PM

12:00-14:30
休息

21:00-24:00
休息（相当于夜间睡眠）

14:30-21:00
摄食(相当于晚餐)

GIANT
PANDA!

大熊猫 201 问

原本是食肉动物的大熊猫，改变了食性，却没有改变肠胃的吸收能力。竹子营养比较低，大熊猫的吸收率也低，因此，它们只能改变生活习惯，用不断地进食来满足生存所需要的能量。

有趣的是，大熊猫即便不吃肉了，也不会委屈自己的口味。吃竹子，它们也会精挑细选，活脱脱一位竹子品鉴大师。无论季节变换，无论何时何地，竹笋永远都是它们的最爱，只因为竹笋的水分和营养物质都十分丰富，特别是糖分含量最高，甜甜的，吃起来十分可口。

一到春天，新笋冒尖的时候，大熊猫就会在竹林里穿巡，采食新笋。而到了夏天，它们爱吃一两年生的竹秆，秋季时喜欢采食竹叶，冬季则爱食老竹笋。大熊猫不仅挑剔竹子是否鲜嫩，还对竹子的分布区域和种类有要求。它们最爱采食海拔在 1800~3200 米范围内的各种竹类植物。

选好方位，选好地点之后，美食家大熊猫还要进一步地分片、分株去选择。这时候，大熊猫会采用"之"字或者倒"8"字形，分别细选，边走边选食。一切准备就绪，大熊猫找到一片空地，专心致志地坐着或者躺着吃了起来。

虽然大熊猫看起来慢吞吞的，但是吃饭的时候，它们毫不含糊。从选食、准备到吃入食物的过程中，没有任何多余的动作，似乎显得很紧凑，迅速地咬切，略加咀嚼，常常在前一口还未吞下时，就又伸出前爪去抓竹子。经过漫长的使用和演化，大熊猫的腕骨（桡侧籽骨）也进化为了一根伪拇指，帮助它牢牢抓住竹子。

大熊猫的演化包含着特征和习性的复杂因素，它们一直针对着环境的变化做着有利生存的改变，唯一不变的是，800万年来，在野外竹林中自由自在的灵魂。

06
与人为邻
Neighbor of Human

要拯救一个物种的最好办法就是保护其所在群落的整体性、稳定性和物种内部遗传的多样性。今天，在保护大熊猫的我们必须固守的最后一块阵地，就是保全那些充满野性的自由生活的种群。

憨态可掬的大熊猫从祖先开始，历经小种大熊猫、巴氏大熊猫，最后发展到现今的现生大熊猫，它曾勇敢地面朝大海，翻山越岭；它曾告别了剑齿虎、剑齿象，最终伴随着人类一路走到了今天，成了生物进化史上最著名的动物"活化石"之一。

如今，大熊猫不仅仅是分布在中国的珍稀动物，也是当今濒危物种保护的代表之一。大熊猫早已与中国的文明和文化融为一体，是中国的"国宝"。

20世纪80年代，中国以及世界范围内许多动物专家，将保护大熊猫的议题放在了紧要的位置上。同时，也涌现出了一大批投身荒野几十年、追寻大熊猫踪迹的研究人员。大熊猫生活的地方常常是在崇山峻岭之中，

它们非常善于隐藏自己的踪迹，因此，野生大熊猫的保护工作有着极大的困难。

对于中国来说，大熊猫具有生态和文化的双重价值；对于投身大熊猫保护的人们来说，大熊猫能否继续繁衍，这个种群是否可以不断壮大，才是他们真正关心的。为此，他们有人付出了一生中的大部分时间，还有人为此献出了生命，依然还有人不遗余力地去荒野中追寻大熊猫。

到今天，人们对大熊猫的认知程度是前所未有的。1999 年到 2003 年，中国完成了第三次大熊猫调查，当时成年大熊猫的数量为 1596 只。到了 2015 年 2 月，国家林业局公布了全国第四次大熊猫调查结果，调查显示 2011 年到 2014 年年底，全国野生大熊猫种群数量达到了 1864 只，相比于前 3 次调查，大熊猫数量正在以合理的种群结构稳定增长。但同时，我们也不得不看到，大熊猫仍然面临总体种群太少、人类干扰、栖息地破碎化、疾病和天敌等多方面的威胁。在继续加强针对野生大熊猫种群及其栖息地保护和研究的同时，迁地保护自然就成为就地保护的一种必要补充手段。科学家们通过不懈努力，成功攻克了圈养大熊猫的繁殖难题，目前圈养大熊猫种群数量已经达到 600 余只，种群遗传多样性不断提高。同时，放归大熊猫的念想很早就在迁地保护工作者心中扎下了根。动物是属于大自然的，自由地生活在山林里的大熊猫才是真正的大熊猫。从 2003 年开始，中国开始了大熊猫的

"野化放归"计划。截至目前，9只人工繁育大熊猫被放归自然并成功融入野生种群。

我们和大熊猫同住一条河，同饮一江水，所以保护大熊猫，不仅保护了它们的栖息地，也保护了人类自己。在中国的秦岭、岷山、邛崃山、大相岭、小相岭和凉山，形成了世界上最为壮观、最为珍贵的野生动物群落。大熊猫、羚牛、金丝猴等动物在这一片栖息地繁衍后代，生生不息，也让这一片土地充满蓬勃生机。从野外追踪到栖息地的保护监测，从迁地保护到野化放归，我们从认知大熊猫，到繁育大熊猫，最终要让其在一片完整、连续和更为广阔的家域永续繁衍。2021年，中国大熊猫国家公园的设立，就是大熊猫保护的呈现终极形式。大熊猫国家公园将见证中国生态文明建设的世界贡献。

The Giant Panda
You may not Know

你不知道的大熊猫

GIANT
PANDA!

GIANT 大熊猫 201 问
PANDA!

Meet the Giant Panda
认识大熊猫

GIANT
PANDA

1. 大熊猫的直系祖先是谁?

答: 始熊猫

Ailurarctos lufengensis。

2. 大熊猫距今约有多少年历史?

答: 从目前的化石记录看,始熊猫已有约 800 万年的历史。

3. 巴氏大熊猫是什么样子的?

答: 根据化石判断,大熊猫的体型演变是由小到大,再变小,巴氏大熊猫是大熊猫演化过程中,体型最大的。比现生大熊猫大 1/9~1/8,其他特征与现生大熊猫相似。

4. 大熊猫家庭在什么时期最兴盛?

答：更新世。

5. 最早记载大熊猫的书籍是什么?

答：在中国古籍中，最早记载大熊猫的书籍是《尚书》和《诗经》。它们距今有近 3000 年的历史。

6. 中国历史上大熊猫的名称有哪些?

答：据记载黄帝时期称貔貅；战国时期称貘白豹、食铁兽；三国时期称貘；西晋时期称驺虞；唐代称白熊；明代称貘。

7. 大熊猫的历史分布情况?

答：在更新世时期，大熊猫进入到一个昌盛年代，广泛分布于我国大部分地区，向北散布到北京周口店一带。此外，东南亚的缅甸、越南、老挝、泰国也有分布。

8. 为什么大熊猫素有"活化石"之称？

答：从迄今为止的发掘情况来看，大熊猫化石的分布涉及我国 14 个省份共 48 个点，其地质年代属于更新世中、晚期，大熊猫是当时的广泛分布种，与其他动物组成了当时具有代表性的动物群落，古生物学家称之为大熊猫剑齿象动物群，由此说明大熊猫在当时居于优势地位，种群繁衍昌盛。长期以来，由于自然环境和气候条件发生过巨大变化，同时期的许多物种因为不能适应这些变故而相继绝灭，变成了化石。经过长期的生存竞争，大熊猫顽强地生存了下来，被称为"活化石"，这是大自然留给我们人类的宝贵财产。

9. 大熊猫在动物分类学中属哪一科？

答：关于大熊猫分类地位的争论已有 130 余年，动物分类学家还存在一定的分歧。有些学者认为大熊猫划归浣熊科或大熊猫亚科。大部分学者认为大熊猫还是与熊接近，但又存在差异。所以现在公认的分类为：

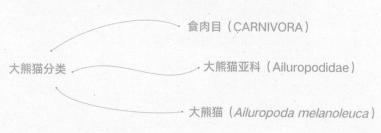

食肉目（CARNIVORA）

大熊猫分类

大熊猫亚科（Ailuropodidae）

大熊猫（*Ailuropoda melanoleuca*）

10. 现在大熊猫分布在哪里？

答: 大熊猫现今分布于陕西的秦岭, 四川、甘肃的岷山, 四川的邛崃山、大相岭、小相岭和凉山六大山系的高山深谷之中。

11. 野生大熊猫在什么海拔范围活动？

答: 野生大熊猫活动范围的海拔主要在 2000~3700 米。海拔 1300 米以下, 由于人为因素影响, 不便于大熊猫活动。活动范围海拔上限在 3200~3700 米, 特殊情况下可达 4000 米。

12. 大熊猫为什么是黑白两色？

答：大熊猫栖息地在 2000~3700 米区域，气候较寒冷，大熊猫的颜色是为了适应自然界的生存规律。专家们推断大熊猫独特的皮毛颜色具有一系列功能，使其能够在不同的环境中匹配其背景，并利用面部特征进行交流。大熊猫面部为乳白色，腰背部为白色；眼周有椭圆形黑色眼圈，耳朵毛色为黑色，前肢至肩以下、后肢的股外侧为黑色。

13. 有彩色大熊猫吗？

答：大熊猫除常见的黑白相间毛色外，还有其他毛色变异。在中国陕西的佛坪国家级自然保护区和长青国家级自然保护区先后发现过棕色大熊猫，在中国四川的卧龙国家级自然保护区发现过白色大熊猫。

14. 大熊猫的尾巴是黑色的还是白色的？

答：白色。

15. 大熊猫的被毛柔软吗？

答：大熊猫被毛比较粗硬，且具有一定弹性，底层有绒毛。它的被毛具有良好的保温防潮作用，所以大熊猫在寒冷的雪地里可以正常生活。

16. 大熊猫的寿命有多长？

答：野生大熊猫的最高寿命为 26 岁，平均寿命为 19 岁。人工饲养的大熊猫寿命在 20~30 岁，目前最高达 38 岁。

1 岁 ≈ 3~4 岁

大熊猫　　　　　人类

17. 大熊猫的年龄阶段是怎样划分的？

答：大熊猫从出生到 0.5 岁龄为婴儿期，
0.5~1.5 岁龄为幼年期，1.5~5 岁龄为亚成体，5~21 岁龄为成年，21 岁龄以上为老年。

0.5 婴儿期

0.5~1.5 幼年期

1.5~5 亚成体

5~21 成年

>21 老年

18. 野生大熊猫的年龄怎样鉴定?

答: 一种方法是根据对大熊猫粪便中的竹子咬节、咀嚼程度和粪团大小对大熊猫进行年龄分组,从而判定其大致年龄。另一种方法是通过牙齿切片,科研人员一般通过看大熊猫牙齿的颜色和磨损程度以及大熊猫的外部形态来判断其年龄大小。

19. 大熊猫的体重是多少?

答: 野生大熊猫的体重 90~110 千克,最重可达120~130 千克;人工饲养的大熊猫体重 90~125 千克,最重可达 180 千克。

20. 大熊猫有多高?

答: 大熊猫肩高 0.65~0.75 米,臀高 0.64~0.65 米。

21. 大熊猫体长是多少?

答: 大熊猫体长一般在 1.2~1.8 米。

肩

臀

头

后肢

前肢

臀

尾

53

22. 雌性、雄性大熊猫体型大小有差异吗？

答：有差异，雄性体型略大，雌性与其相差约
10%~18%。

23. 大熊猫的雌、雄性在数量上占比是
多少？

答：大熊猫的雌、雄性数量比约为 1.16∶1。

24. 大熊猫的足有多长？

答：大熊猫的足长 12~20 厘米。

25. 大熊猫的脚有几趾？

答：大熊猫的前、后脚各有 5 趾，其前脚还有一个由
桡侧籽骨特化成的伪拇指。这个"伪拇指"可帮助它
们紧紧抓住竹茎，灵活地采食竹子。

26. 大熊猫初生幼仔尾巴怎么会那么长?

答: 大熊猫的祖先是食肉动物, 幼仔刚出生到成年时携带了更多它们祖先的特征——长尾巴。在大熊猫从小到大的生长发育过程中, 尾巴会慢慢停止生长, 成年后的大熊猫尾巴就相对短了许多。

27. 为什么大熊猫长大后尾巴变短了呢?

答: 从出生到成年, 大熊猫的尾巴相对于身体来讲, 长度增长得极少, 主要是宽度有所增加。相对于成年大熊猫的体长和庞大的体型而言, 扁平的尾巴就不太突出了, 以致造成很多人的误解, 以为大熊猫没有尾巴。

28. 大熊猫宽短的尾巴有何用处?

答: 大熊猫的尾巴比较短, 长度约 10~12 厘米, 毛茸茸的, 通常贴着身体。当尾腺、肛周腺和外阴有分泌物泌出时, 大熊猫可以把尾巴当作刷子, 四处标记信息。

29. 大熊猫经常做标记的目的是什么?

答: 每只大熊猫都会在自己领地边缘的树木或石头上用肛周腺标记气味,以此警告侵入者禁止入内。大熊猫做标记的目的是告诉其他大熊猫自己的生理信息,并通过嗅闻其他大熊猫的气味标记,解读其他大熊猫是否处于发情期等信息。

30. 古今大熊猫的体貌和器官有哪些变化?

答: 从大熊猫的生理解剖来看,无论其外部形态,还是部分器官构造,都具有原始特征和适应性的变化特点。最明显的是为了适应主食竹子,从头骨、牙齿与四肢到消化道都发生了一些变化。器官变化主要是臼齿咀嚼面变大和进化出的第六指。

31. 大熊猫正常体温是多少度?

答: 大熊猫的正常体温在 36.5~37.5℃。

32. 大熊猫正常呼吸频率是多少?

答: 大熊猫正常呼吸频率 16~24 次 / 分钟。

33. 大熊猫的正常心跳是多少？

答：大熊猫的正常心跳为每分钟 70 次。

34. 大熊猫有多少对染色体？

答：大熊猫有 21 对染色体，其中 20 对为常染色体，1 对为性染色体。

35. 大熊猫的脑容量有多少？

答：大熊猫的头看起来大，但它的脑并不发达，其脑容量只有 310~320 毫升，比野熊的脑容量还少 60 毫升。

36. 大熊猫有多少枚牙齿？

答：大熊猫乳齿有 24 枚，恒齿 40~42 枚。

37. 大熊猫有多少枚椎骨？

答：大熊猫共有 42 或 43 枚椎骨。

38. 大熊猫有多少对肋骨？

答：大熊猫的肋骨左右不对称，左边 13 根，右边 14 根或左边 14 根，右边 13 根。

39. 大熊猫有多少对脑神经？

答：大熊猫有 12 对脑神经。依次为：嗅神经、视神经、动眼神经、滑车神经、三叉神经、外展神经、面神经、听神经、舌咽神经、迷走神经、副神经、舌下神经。

40. 大熊猫有多少对脊神经？

答：大熊猫有 34~35 对脊神经。

41. 大熊猫的脸型有区别吗？

答：据不同的大熊猫个体，可简单地分为圆脸和长脸。个别大熊猫的耳朵和眼圈有大小变化。目前，有关研究人员已经建立起了大熊猫面部识别系统，可以识别录入大熊猫谱系的任何个体。

99%

42. 大熊猫是怎样适应环境变化的？

答：大熊猫选择了适者生存的策略，本来是食肉动物的身体结构和器官发生了适应性退化，以适应主食竹子的客观条件。

43. 大熊猫是素食者吗？

答：在动物分类地位中，大熊猫被划在食肉动物类群里，但大熊猫可以说是一个典型的食物特化者，主要是素食。在野外环境里，大熊猫的食物里只有约 1% 的部分会吃竹子以外的其他植物与动物；在人工饲养环境里，大熊猫还能取食蜂蜜、鸟蛋、山芋、灌木叶、柑梅、香蕉等。

44. 大熊猫的食谱是怎样的？

答：大熊猫主要以竹子和竹笋为食，约占食物总量的 99%，其中被吸收的仅 17% 左右。在人工饲养环境下，为了保证大熊猫

身体发育的营养需求，还会提供适量的奶、苹果、胡萝卜和精饲料。在野外，大熊猫偶尔也会吃一些其他植物和动物打打牙祭，所以大熊猫是一种以竹子为主食的杂食性动物。大熊猫嗜水，在野外只爱饮用流动水，每次饮用足量，有时还会因为饮水过多而导致行走困难。大熊猫的饮水习惯有助于它们对竹子的消化。

45. 野生大熊猫每天花多长时间吃竹子？

答：根据研究，大熊猫在野外每天花费10~14 个小时取食竹子。

46. 大熊猫一天要吃多少竹子？

答：大熊猫生活在高寒地区，4~6 月为春季，7~10 月为夏秋季， 11 月至来年 3 月为冬季。春季大熊猫以新竹笋为食，在这个季节的食量也最大， 一般自取食 30~60 千克，平均约 40 千克。到了夏季，食物变成以竹茎为主， 还兼食一些较老的竹笋和新发的幼枝叶。秋季，是当年幼枝叶最繁茂时期，它们的食谱以幼枝叶为主。这时的嫩枝幼叶营养成分居竹的其他部分之首，而所含的粗纤维又低于竹茎。因此，在这个季节大熊猫每天只需采食 10~14 千克，而且活动范围达到一年中最小，也节省了活动消耗的能量。

47. 野生大熊猫为什么要吃大量的竹子？

答：大熊猫虽然取食竹子，但其消化道是典型的食肉动物特征，消化道短而缺乏黏膜，没有盲肠，因此，不能很好地消化竹子细胞壁中的纤维素和木质素等成分，只能消化吸收细胞内含物和部分半纤维素。所以，不像其他食草动物具有复杂反刍胃或很大的盲肠，能够充分地消化和吸收取食的竹子。为了满足营养需要，大熊猫每天花很多的时间大量取食竹子。

48. 大熊猫最喜欢吃哪种竹子?

答: 大熊猫除了最优化组成季节食谱外, 在摄食行为上也采取最优化选择。吃竹笋季节, 它们会因地择优。每一个山系的大熊猫都有一种最喜欢吃的竹子, 邛崃山系的大熊猫最喜欢的是冷箭竹, 冷箭竹较细, 大熊猫不食其幼笋, 只吃冬季的老笋。秦岭山系大熊猫, 主要采食巴山木竹, 这种竹的笋较粗, 大熊猫择食直径 12 毫米以上的竹笋。在其他山系, 大熊猫主要择食缺苞箭竹、拐棍竹和玉山竹等, 择食直径 9 毫米以上的竹笋。在大熊猫生活的森林里, 有很多种竹子, 大熊猫能够采食十几种竹子。

49. 大熊猫吃竹子有哪些讲究?

答: 大熊猫在进食竹子时, 会将竹叶收集成把, 一口一口地吃; 吃竹竿时, 大熊猫会用牙齿剥掉竹子外皮, 食用内竿。根据季节不同大熊猫喜食的竹子种类和部位也有差异。

50. 大熊猫消化道的构造是怎样的?

答: 大熊猫消化道和其他哺乳动物一样, 由口腔、食道、胃、小肠、大肠构成。

51. 大熊猫的肠有多长？

答：大熊猫的肠分为小肠和大肠，小肠的平均长度为490厘米（269~624厘米），大肠的平均长度为120厘米（72~167厘米）。

52. 大熊猫的肠与草食动物有何不同？

答：一是大熊猫整个肠道短，约为自身体长的4.3倍；而草食动物消化道长，约为自身体长的10~30倍。二是大熊猫的结肠与回肠直接相连，无盲肠，结肠平直，无纵带和肠袋。三是大熊猫的肠道中有丰富的黏液腺，而草食动物却没有。

53. 大熊猫的胃是什么样的？

答：大熊猫的胃属于单室有腺胃，呈典型的"U"字形曲囊袋状。

54. 坚硬的竹子会刺伤大熊猫胃肠吗？

答：不会。因为大熊猫消化道的肌层厚，消化道中存在丰富的多细胞黏液腺，这些黏液腺分泌的黏液对消化道起着保护作用。

55. 大熊猫每天要排出多少粪便？

答：据统计，成年大熊猫每天排便 40 多次，排出的粪便在 120 团左右，若吃竹笋，排出的粪便为 150 团，每团粪便的重量约 100 克，每天排出粪便的总量约为 15~20 千克。

56. 大熊猫一天的作息时间是怎样的？

答：0:00-2:00 休息；2:00-7:00 摄食（相当于早餐）；7: 00-10: 30 休息；10: 30-11:00 漫游、玩耍；11: 00-12: 00 摄食（相当于午餐）；12:00-14:30 休息；14:30-21:00 摄食（相当于晚餐）；21:00-2:00 休息（相当于夜间睡觉）。

57. 大熊猫只在白天活动吗？

答：大熊猫不只是白天活动，它在夜晚也能活动，属昼夜兼行动物。

58. 光照对大熊猫有什么影响吗？

答：适当的光照对大熊猫的生长发育具有积极的影响，光照还能左右大熊猫的发情时间及发情过程的长短。但夏秋季节强烈的光照应尽量避免，特别是离开大熊猫栖息地环境后的大熊猫，如果环境温度过高、紫外线强烈，都会导致大熊猫中暑或患热射病，这种情况往往会造成不可逆转的伤害。

59. 大熊猫冬眠吗？

答：大熊猫没有冬眠的习性。由于竹子是四季常绿植物，大熊猫冬季依然可以取食竹子过冬。冬季，大熊猫通常从高海拔迁移到中低海拔区域，低海拔地区气温相对较高，竹子等食物相对充足。

60. 大熊猫在哪里睡觉？

答：在野外，出生不久的大熊猫幼仔待在妈妈为它们准备的温暖树洞或岩洞中；长大一些后，它们最喜欢在树上睡觉，这样能够躲避天敌的袭击；成年大熊猫总是走到哪里就在哪里休息，但它们最喜欢选择一个可以倚靠的地方，比如大树边或倒木旁。

61. 大熊猫有哪些睡姿？

答：说起大熊猫的睡姿，可以说是多种多样。大熊猫自幼仔到成年，一般卧睡、仰睡较多。大熊猫的卧睡姿势很有意思，无论在栖架上、睡床上，还是在树杈上，四肢都会自然下垂，有一种倍感松弛的感觉。仰睡时，一般后肢会搭在一个固定物上，很多时候还用前肢捂着眼，很像我们用胳膊挡着眼睛遮光，十分滑稽。大熊猫躺着的时候，也会伸懒腰，并时而打哈欠。

62. 大熊猫的生活习性有哪些？

答：大熊猫长期生活在竹林茂密的森林里，喜欢独来独往，它们通常用尿液和肛周腺分泌物来标记自己的领地范围。大熊猫喜冷怕热、擅长爬树、会游泳、不冬眠。一旦发现异常情况，就会迅速在竹林中躲避或爬上树梢。

63. 大熊猫怕冷还是怕热？

答：大熊猫惧酷热，不畏寒冷。

64. 大熊猫会游泳吗？

答：大熊猫可以说是游泳高手。当天气炎热时，它们就可以趟河、戏水、游泳。大熊猫也可以长距离深水游泳，这也是它们在复杂环境里得以迁徙的主要手段。

65. 大熊猫会爬树吗？

答：会，它们是爬树高手。当生下来的大熊猫幼仔长到六个月大时，就可以跟随妈妈爬树了，在树上生活和睡觉。

66. 大熊猫趴在树上的时间多吗？

答：大熊猫半岁以后就喜欢攀爬，爱爬树是其生存的本能。特别是在野外出生的大熊猫幼仔，雌性大熊猫要出去采食竹子，在离开母亲保护的时间里，幼仔会受到一些掠食动物的威胁，因此幼仔会迅速爬上树以躲避天敌的侵害。另外，趴在树上容易晒到阳光，使身体得以正常健康发育。

67. 大熊猫喜欢水吗？

答：大熊猫喜欢饮水，也喜欢戏水。在野外只爱饮用流动水，如溪水或源头水。若结了冰，大熊猫会用前掌击破冰层；若水浅，会挖成水坑饮水。有时大熊猫饮水量过大，腹胀如鼓，行似醉汉，走路蹒跚或躺在水边，民间说它有"醉水"的习性。尤其是在夏天，大熊猫还会在小溪浅出泡泡脚或坐着梳理毛，有时还会用前肢往身上刨水。

68. 大熊猫走路的姿势是什么样的？

答：大熊猫走路是内八字姿势。内"八"字步有利于在竹林中穿梭、跑动。大熊猫以内"八"字步行走或跑动，这也是适应森林环境的一项重要变化。

69. 大熊猫会站立吗？

答：大熊猫站立的功夫十分了得，它会站立取食高处的东西，站立时还会张望打探，谨慎观察周边动静。厉害的是，它还会倒立，一般是倒立在树干上做肛周腺的气味标记。另外，与发情雌性大熊猫交配时，雄性大熊猫有时也要站立交配。

70. 大熊猫为何喜欢做翻滚动作？

答：大熊猫特别高兴的时候就喜欢做翻滚动作，前滚翻、后滚翻都可以。做这些动作时，多为春季阳光明媚的早晨，或秋季凉爽的时候，还有就是大雪之后，在雪地里兴奋地打滚。有时还会仰躺起身体在原地转圈。大熊猫，特别是幼仔对外界的事物具有强烈的好奇心，因此，它边玩、边滚、边探索，通过滚爬感知周边事物。

71. 大熊猫也要梳理被毛吗?

答:大熊猫如果戏水以后,会左右甩动身体,把被毛上的水甩掉。吃完东西也要舔舐嘴的周边及前掌。不时用前爪梳理耳朵和颈部,还会用后肢挠痒。够不到背部时,它就靠在树干或石头上摩擦,起到挠痒和清理被毛的作用。

72. 大熊猫为什么不喜欢人摸它的耳朵?

答:因为大熊猫的耳朵被毛薄,听觉很灵敏,当触摸它们的耳朵时,它们很敏感和警觉,感到不适。若反复这样做,它们会感到心烦而出现进攻行为。

73. 大熊猫的听觉如何?

答:大熊猫的听觉很灵敏。能听到很远的地方传来的声音。

74. 大熊猫的嗅觉如何?

答:大熊猫的嗅觉非常敏锐,通过敏锐的嗅觉,大熊猫识别出其他个体发出的气味信息。

75. 大熊猫的视觉如何？

答：长期生活在深山密林之中的大熊猫，虽然看得不远，但天生不是近视眼，视力功能和它的听力以及嗅觉能力相差太远了。

76. 大熊猫的叫声是什么样的？

答：大熊猫在不同的境况、不同的情绪和心理状态下的叫声不同。主要有：牛叫声、羊叫声、呻吟声、吠声、唧唧声（幼犬鸣叫），还有嗷嗷叫、嗥叫、鸟叫、吱吱叫等。

77. 大熊猫会伤人吗？

答：一般说来，大熊猫性情温和，不主动进攻其他动物。但在遇到危险的情况下也会伤人。特别是在发情求偶和产仔季节，进攻性强，更容易伤人。

78. 大熊猫会生什么病？

答：大熊猫也会生病。通常分内科疾病、外科疾病、产科疾病、传染疾病和寄生虫病几大类。

GIANT
PANDA! 大熊猫 201 问

79. 野生大熊猫最常见的疾病是什么？

答：寄生虫病、胃肠道疾病。

80. 大熊猫也会有紧张和压力感吗？

答：有，特别是在人工饲养条件下，人为的噪声或周边的干扰。如果持续不断地作用于大熊猫，就会增大它的紧张和压力感。

81. 大熊猫也会患感冒吗？

答：感冒是一种由于气温变化，导致大熊猫身体免疫力下降和多种相关病毒入侵而引起的上呼吸道症状为主的疾病。明显症状是咳嗽、流鼻涕、发烧，多数精神萎靡不振，卷睡。一般肌肉注射安乃近、柴胡，也可口服泰诺林。人类感冒也会传染大熊猫，因此，身患感冒的工作人员须待病情痊愈后才能继续从事大熊猫的饲养工作。

82. 大熊猫也会中暑吗？

答：大熊猫的野外栖息地基本处于海拔 2000 米以上的区域，气温较低，常年平均气温为 10~15℃。其居住环境也是高山森林地带，很多时候是云雾缭绕，不会受到阳光的直射。但在人工饲养环境下，就必须注意大熊猫的生活环境温度与光照强度。特别是在常年气温偏高、光照强烈的地区，必须考虑避光、降温与通风问题。如果出现重度中暑，需要紧急采取降温措施，否则，大熊猫会因为中暑而导致死亡。

83. 大熊猫是群居生活吗？

答：不，它们属于独居动物，在野外都有属于自己的领地，不允许其他大熊猫侵犯，否则会出现打斗行为。

84. 大熊猫总是处于游荡状态吗？

答：大熊猫居无定所，不固定采食竹子的地点，大便也是随地而排，总是在自己的领域内游荡。大熊猫根据竹子的取食环境和水源而随机移动、休息，这大大节约了外出觅食和返巢的时间与能耗。

85. 大熊猫在野外有固定的栖息处所吗?

答: 大熊猫在野外没有固定的栖息处所,它们边走边吃,喜单独活动,四处游荡。它们不会筑窝,常栖于树洞、岩穴或树下乱草堆处,多早晚出来觅食和饮水,其余时间多是睡眠。它们的视觉距离较短,行动也较缓慢,但却能快速而灵活地爬上高大的树木,而且能泅渡湍流的河溪。

86. 大熊猫如何求偶?

答: 每年的 3~5 月是大熊猫的发情季节。在野外,雌性大熊猫在发情季节里通过叫声和标记物的特殊气味,将居住在领地周围的雄性大熊猫纷纷吸引到一块。通常雄性大熊猫都要经过几天的斗争来取得与雌性大熊猫的交配权。雌性大熊猫最佳受孕时间通常只有 1~2 天。交配完成后,雌性大熊猫会将雄性大熊猫赶出自己的领地,继续自己的独居生活,直到宝宝出生。

87. 大熊猫一般在什么季节婚配?

答: 大熊猫的发情, 时间多在春季, 各地略有差异, 栖息地的报春花开时, 正是它们的发情期。一般在 3 月下旬开始, 一直延至 5 月中旬, 多在 4 月中旬, 个别也有在 1~9 月。发情周期是两年一次。此外, 通常大熊猫发情期也受到栖息地的纬度、海拔和气候的影响。

88. 自然环境中的大熊猫是一夫一妻吗?

答: 大熊猫不是一夫一妻。

89. 大熊猫在什么季节产仔?

答: 一般情况下, 大熊猫于春季发情交配, 秋季产仔, 多集中在 8~9 月。极个别个体在第二年春季产仔, 目前仅在四川卧龙有一例。

90. 大熊猫如何哺育后代?

答: 大熊猫的怀孕期通常为 3~5 个月, 最长可达 10 个月。野生大熊猫一般会选择比较隐蔽的岩石洞或树洞作为产房。初生的幼仔非常脆弱, 需要一直待在妈妈怀里, 母

乳是它们唯一的营养来源。幼仔 4 个月大时才学会走路，1 岁左右才能吃竹子，所以一般大熊猫幼仔都会跟随妈妈一起生活到 1~2 岁左右，之后才会离开妈妈独立生存。

91. 大熊猫的身体变化是怎样的？

答：大熊猫幼仔出生时身体呈肉红色，体重一般在 80~200 克，部分身体器官尚未发育完全，皮肤表面有稀疏的白色被毛，像一只小老鼠。7 天后，四肢、耳朵和眼睛周围渐渐出现黑色。20 天左右，黑白相间的体毛长齐，皮肤具备调节体温的能力。40 多天眼睛才会慢慢睁开，2 个月时能完全张开，但此时还是无法站立行走。3 个月的大熊猫幼仔已经和我们见到的大熊猫一模一样了，它们开始学走路。

92. 大熊猫通常一胎生几只宝宝？

答：在野外环境中，大熊猫一般每胎生一只幼仔，但在人工饲养环境，50% 的大熊猫大多每胎能生 2 只幼仔，极少生 3 只幼仔。

93. 雌性大熊猫一生最多能生多少宝宝？

答：在野外环境中，雌性大熊猫一生中能生产 4~8 只幼仔。

94. 首次发现野生大熊猫双胞胎是在何时何地？

答：1990 年在四川省卧龙国家级自然保护区内首次发现已育幼 45 天左右的野生大熊猫双胞胎幼仔。

95. 雌性大熊猫怀孕时间有多长？

答：大熊猫有延迟着床现象，最短 73 天，最长 324 天，一般平均在 120 天左右。

96. 为什么大熊猫的妊娠时间差异如此大？

答：因为大熊猫的受精卵有延迟着床现象。雌性大熊猫在春季发情配种后，受精卵从输卵管到达子宫并没有立即着床，而是在子宫中游离一段时间后再着床，延迟着床时间在 1.5~10 个月。

97. 野生大熊猫在什么地方产仔?

答：野生大熊猫在临产前开始寻找适宜的洞穴营巢。在原始森林里，它们选择古老的空心树洞或大树根际处的洞穴。在采伐过的次生林中，缺少古老大树，它们便选择林中的天然石洞或石穴。

98. 刚出生的大熊猫宝宝是什么样子?

答：刚刚出生的大熊猫幼仔像只裸露的老鼠，有个大大的脑袋和长长的尾巴。体长15~17厘米，尾长4.5~5.2厘米，后足长2.2~2.5厘米，体重为100克左右，两眼紧闭，全身粉红色，有稀疏的白毛，耳似两颗肉粒状，不能站立和爬行，但能抬头啼叫，且叫声洪亮。

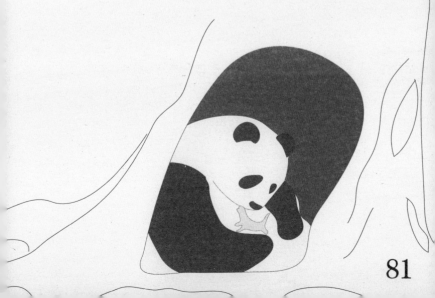

99. 新出生的大熊猫宝宝吃什么?

答: 新出生的大熊猫幼仔只吃妈妈的奶水。幼仔出生的 14 天内母亲寸步不离，常在母亲怀里发出尖厉的叫声。第二年春天，幼仔已经开始随母亲行走，开始品尝幼嫩的竹叶，母亲不在身边的时候，它通常是在树上度过，高大的针叶树成为它躲避天敌的安全岛。出生 8~9 个月，幼仔开始被断奶，开始跟母亲学习在不同季节、不同地形采食的本领。

100. 新生的大熊猫宝宝生长得快吗?

答: 新出生的大熊猫幼仔很小，体重只有母亲的 1/1000，但新出生的大熊猫幼仔生长比较迅速，3 个月大的幼仔能有 5~6 千克，6 个月大时 12 千克左右，1~4 岁的幼仔重量可以分别达 38、72、87、97 千克。

101. 幼龄大熊猫日平均增重多少?

答: 幼仔出生后，由于自身失水及对新环境的不适应和哺乳的原因，前 3 天的体重一直低于出生时的体重，到第 4 天才能恢复到出生时的体重，从第 5 天开始，随着日龄的增长，体重近似直线增长。1 月龄时平均体重将达到 1.2 千克左右，约为初生时体重的 10 倍，由

此可见，幼仔的生长发育速度是非常快的。此后，第 1 个月每日平均增重约 36 克，第 2 至第 6 个月平均增重 74~80 克。

102. 幼龄大熊猫体长的增长情况是怎样的？

答：体长变化是反映幼龄大熊猫生长发育状况的一项重要指标，正常情况下，幼仔的体长与日龄呈直线递增的关系。大体而言，初生第 1 天为 15 厘米，30 天为 30.1 厘米，2 月龄为 42.1 厘米，3 月龄为 53.2 厘米，4 月龄为 63.0 厘米，5 月龄为 70.5 厘米，6 月龄为 80.3 厘米。

103. 大熊猫妈妈有几种育幼姿势？

答：初生幼仔降临后的前三天，大熊猫妈妈主要以坐姿哺育幼仔。随着幼仔的长大，仰卧、侧卧育幼姿势增多，搂抱的力度也随之下降，渐渐给幼仔一定的活动空间。幼仔刚出生的几天，大熊猫妈妈一般都尽量搂抱幼仔，很少去进食和排便。即使偶尔离开，时间也很短暂。

大熊猫幼仔出生时身体呈肉红色，体重一般在80~200克，大约相当于妈妈的1/1000。

104. 大熊猫妈妈舔舐幼仔全身有何作用？

答：主要有五个方面的作用：一是清洁全身；二是因为其唾液中含有铁合蛋白，通过细细舔舐，将唾液涂满幼仔全身，具有抗病消毒、防止感染的作用；三是促进幼仔微循环；四是保持皮肤湿度，防止水分超量蒸发；五是刺激幼仔排便。

105. 大熊猫幼仔的早期社交活动有何重要性？

答：经过科研人员几十年来的观察、研究，发现大熊猫幼仔的早期社交活动对成年后的繁殖行为、母性行为会产生巨大影响。由于对圈养大熊猫存在着追求种群数量增加的需求，所以会在幼仔半岁的时候人为与大熊猫妈妈分开，以利于大熊猫妈妈在第二年发情配种。然而，这种做法会让半岁的幼仔很早就失去了与大熊猫妈妈一起生活和学习交流的机会，有些幼仔甚至与其他幼仔交流的机会也没有。缺失了早期社交活动的大熊猫，成年后多半会出现很多不正常的行为，比如强烈的攻击性、不正确的交配姿势、对初生幼仔哺育方式不正确等。相对而言，幼仔早期社交活动的缺失，对雄性大熊猫的不利影响更大、更显著。

106. 大熊猫初生幼仔与大熊猫妈妈有哪些交流方式？

答：大熊猫初生幼仔在不具备听力与视力前，主要依靠触觉、嗅觉及叫声与大熊猫妈妈交流，反馈冷热、饥饿、捂抱的松紧、寻找乳头等信息。大熊猫妈妈会根据幼仔的叫声来判断其不同的需求，有时会挪动幼仔靠近乳头，有时会舔舐幼仔，个别时候还把幼仔叼在嘴里四处走动。总而言之，大熊猫妈妈会根据幼仔的需求随时调整自己的行为方式，所以，在生下幼仔后的前三天时间里，大熊猫妈妈会非常辛苦。

107. 初生大熊猫幼仔有哪些行为？

答：初生大熊猫幼仔在母体内发育很不完善，其感官系统尚未完全建立，幼仔与母体之间主要依靠声音进行交流。其主要行为有：

（1）尖叫 这种叫声分为两种情况，一是声音较低，一是声音很高。当幼仔露出母体外感到寒冷时发出这种叫声，通常是要求大熊猫妈妈调整体位。

（2）呱呱叫 这种叫声类似蛙鸣，但声调很低。此叫声多紧接尖叫之后发出，表示一种愉快的感觉，是大熊猫妈妈调整体位后幼仔感到舒适时发出的信号。

（3）连续尖叫 这是幼仔感到极端不适或要求大熊猫妈妈哺乳时发出的信号，有时持续时间较长，直到幼仔感觉满意方才停止。

（4）颤抖 当幼仔露出母体外时，幼仔会因为感到寒冷而致身体颤抖，此时，大熊猫妈妈会调整幼仔的位置并将其完全遮住以保暖。

（5）吸乳 幼仔感到饥饿时，便会寻找奶头并发出连续的尖叫声，此时，大熊猫妈妈会帮助幼仔寻找奶头。1~5 日龄以前，幼仔吃奶时间间隔不等，每昼夜 6~12 次。幼仔每次吃奶的持续时间也不等，可从半分钟到 10 多分钟，有的长达 30 分钟。15 日龄以后，吃奶次数逐渐减少，降到每天 3~4 次。

（6）嗅闻 随着幼仔的成长，它开始以嗅闻的方式对周围环境进行探究，母子间的交流也开始以这种方式进行。

（7）爬动 2 月龄以前的幼仔几乎没有任何活动能力，除了吃奶就是睡觉，寻找奶头时只能蠕动。3 月龄以后的幼仔能够爬动，但四肢不协调，且爬动距离短。

（8）不平稳走动 4 月龄左右，幼仔开始能够走动，但四肢力量不均衡，走动不平稳，时常翻滚在地。

（9）走动和攀爬 当幼仔四肢力量均衡后，开始正常活动，此时的幼仔不但四处走动而且开始攀爬，但活

GIANT
PANDA!

大熊猫 201 问

动范围不大。有时活动范围稍大时，大熊猫妈妈会将幼仔叼回。

108. 据说大熊猫妈妈会遗弃其幼仔，真是这样吗？

答：雌性大熊猫在一胎产下 1 只幼仔时，它们都会对自己产下的幼仔精心养育；当它们一胎产下两只或两只以上幼仔时，几乎所有的雌性大熊猫都会选择其中最健康强壮的 1 只哺育，而将其余的幼仔丢弃不管，这就是大熊猫的弃仔行为。在饲养情况下，一些雌性大熊猫没有能力照料它们的所有幼仔，被遗弃的幼仔通常给予人工喂养。

109. 近亲繁殖是否影响到野外大熊猫种群的生存？

答：所有物种都发展出了自身独特的避免近亲交配的机制与策略。目前为止，还没有直接的证据显示近亲交配对野生大熊猫的影响来说是个问题。但事实是，在某些区域，大熊猫栖息地破碎化严重，大熊猫被割裂成多个小种群，近亲繁殖导致大熊猫小种群的遗传多样性降低。未来的工作重点是将这些相互隔离的种群相互联系起来，互相交流，提高遗传多样性。

110. 大熊猫出生后多长时间开始长牙？乳牙什么时候长齐？

答：大熊猫出生后 3 个月开始长牙；到 6 个月时，乳牙基本长齐。

111. 大熊猫什么时候开始换牙？多长时间换牙结束？

答：幼仔长到 8 月龄左右乳齿逐渐脱落，开始换牙；15~17 个月乳牙全部被恒齿代替。

112. 大熊猫幼仔什么时候断奶？

答：在野外，从出生到 7 个月以前，幼仔的营养获得全依赖于母体哺乳，到 8~9 个月才开始断奶。

113. 雄性大熊猫会帮助抚育幼仔吗？

答：雄性大熊猫不会帮助雌性大熊猫抚育幼仔，它在交配结束后便离开雌性大熊猫，怀孕、产仔、育幼及幼仔的生存培训等全由雌性大熊猫承担。

114. 野生大熊猫幼仔多大才离开母亲独立生活?

答：野生大熊猫幼仔 1.5 岁时离开其母亲独立生活，女儿被远嫁他乡，但儿子仍然在其母亲的巢域内活动，到 2.5 岁时才开始离开母亲的巢域，去建立自己的领地。

115. 大熊猫幼仔后天学习的重要性有哪些?

答：伴随着自身的成长，大熊猫幼仔会从母亲那里学习到很多本领，如躲避天敌、爬树、选择适口的竹子、寻找水源等。在野外，1.5~2.5 岁就要独立面临生活。一般来说，幼仔长大以后就得独自觅食、玩耍、休息。但在独立生活初期，幼仔通常不会离大熊猫妈妈巢域太远，直到逐渐适应后才会建立自己的活动区域。

0~7 个月
全依赖于母体哺乳

116. 野外能容易看到大熊猫吗？

答：大熊猫被称为"林中隐士"。由于大熊猫生活在茂盛的浓密竹林里，多单独活动，所以在野外很不容易看到大熊猫个体。但如果你进入了大熊猫生活的森林区域，就比较容易能够看到大熊猫生活的痕迹，如它们的粪便和咬断的竹节。

117. 大熊猫栖息地还有哪些同域物种或伴生动物？

答：大熊猫的伴生动物有金丝猴、扭角羚、小熊猫、黑熊、野猪、绿尾虹雉、红腹角雉、林麝、竹鼠、猕猴、毛冠鹿、牛羚、血雉鸡、红腹锦鸡等，这些珍贵动物基本上都与大熊猫生活在同一环境中。它们有着共同的经历，曾经受过冰川的袭击和严峻考验，然而它们都有效地借助于我国西南高山深谷的有利地形而幸存至今。它们通过长期共存，协调地共同生活于同一区域，但各自占有自己的空间，组成了一个较为稳定的动物群落。

118. 大熊猫天敌主要有哪些？

答：大熊猫一般都能与其他野生动物和睦相处，但体弱病幼的大熊猫容易受到豹、豺、狼等食肉动物的危害。

119. 大熊猫应该叫"大熊猫"还是"大猫熊"？

答：大熊猫更应该叫"大猫熊"而不是大熊猫，因为它们是熊类的近亲，却是猫类的远亲。只是20世纪中国大陆一次汉语读写顺序的"西化"改变，造成了这一"谬误"，现代中国人也就认可了这一将错就错的习惯，使这一"似熊非熊"的动物有了不同于其"族谱"的名称连最挑剔的动物分类学家也认可了这一"错误"，因为它们是唯一的、独有的、特别的……在我国台湾省的书刊上仍保着"猫熊"的称呼。

120. 大熊猫的脚掌与熊掌一样吗？

答：大熊猫的前掌跟其他熊类不同，黑熊的前掌相对窄长，中指也长，并前突，爪长而尖。大熊猫的前掌稍圆，五指基本平齐，指缝间有毛。

121. 大熊猫与黑熊到底有哪些区别？

答：黑熊体形修长，全黑色被毛，仅胸前有"一"字形白毛，前掌细长，其爪锐利，头部偏长，吻也长。黑熊是杂食性动物，染色体37对。大熊猫则体型圆润，四肢、耳朵和眼圈呈黑色，其他部位白色，前掌椭圆形、五指整齐，有第六"伪拇指"。大熊猫头部圆形，吻短。大熊猫的食物99％为竹子，前臼齿大而平，染色体21对。

122. 大熊猫头骨部位与其他熊类有何不同？

答：大熊猫是高度特化的食肉动物，其外表虽与熊类接近，但很多解剖学特征又与熊类不同。大熊猫的头骨与相关肌肉明显与其他熊类不一样，咀嚼肌特别发达，前臼齿磨面变大，这些变化都与高度依赖进食竹子有关。

123. 大熊猫的齿型与其他熊类有哪些不同？

答：从解剖学角度看，大熊猫的齿型不同于其他杂食性的熊类，大熊猫的前臼齿平而宽，可以轻松咬断竹子和碾磨碎竹子。

124. 大熊猫与小熊猫是两个不同的物种吗？

答：大熊猫与小熊猫是不同的两个物种。1825 年 6 月，法国人弗雷德里克（F.Georges）在喜马拉雅山地区发现了小熊猫，首先命名为"*Ailurus fulgens*"，英文为"Panda"。因其后来发现了大熊猫的存在，所以英

文称 Lesser Panda 或 Red Panda 加以区分，中文名翻译为小熊猫，Giant Panda 为大熊猫。小熊猫外貌十分漂亮，有狐狸大小，被毛棕红色，脸圆，耳缘、面颊、眼部、嘴部为白色，长而蓬松的尾毛有九个白色的环，当地人称小熊猫为"九节狼"。

小熊猫分布在尼泊尔与我国青藏高原，一部分与大熊猫栖息地重叠。小熊猫喜欢在树上活动，觅食时下到地面，也喜欢采食竹叶。另外，还以野果与鸟蛋、小动物等为食。

125. 大熊猫有很多拟人的动作吗？

答：大熊猫的可爱之处多在于具有很多与人类共通的行为与习性，比如大熊猫也会伸懒腰，时常坐立玩耍。如果视线被挡或需要获得高处物品的时候，它也会站立起来。睡觉时，如果光线较强，它会用前掌盖住眼睛，然后小憩一会儿。睡醒后也会打一个或多个哈欠，表现出非常惬意的神态。特别是吃竹笋和窝头、水果等食物的声音，"吧唧……吧唧……吧唧……"，跟人吃到特别香的食物一样。吃竹秆时，"清、脆、爽"的感觉明显，会诱使人们口水直流。

大熊猫的"伪拇指"可帮助它们紧紧抓住竹茎，灵活地采食竹子。

Giant Pandas' Home in the Wild

大熊猫的野外家园

GIANT
PANDA!

大熊猫一般生活在海拔 2000~3700 米 的针阔叶混交林和亚高山暗针叶林内，林层下生长有茂盛的竹林。

126. 大熊猫的模式标本产地在哪里？

答：大熊猫的模式标本产地在中国四川省宝兴县的邓池沟。

127. 大熊猫过去的家园在哪里？

答：在历史上，大熊猫曾经生活在包括大半个中国，老挝，缅甸北部，泰国北部，越南北部的亚热带常绿林中。由于人类开垦、采伐以及公路建设等原因导致生境大量丧失与片段化，大熊猫种群已经退缩到中国陕西秦岭南麓、甘肃以及四川盆地西缘的岷山、邛崃山、凉山、大相岭、小相岭六个高大山系中。

128. 大熊猫的栖息地面积有多大？

答：2015 年公布的第四次全国大熊猫调查结果显示，野生大熊猫分布在四川、陕西、甘肃三省的 17 个市（州）、49 个县（市、区），栖息地总面积扩大至258 万公顷，野生种群数量增长至 1864 只。因自然

隔离和人为干扰等因素的影响，大熊猫野外种群被分割成 33 个局域种群，其中 18 个种群数量少于 10 只，具有很高的灭绝风险。

129. 大熊猫的野外家园是怎样的？

答：大熊猫属温带动物，一般生活在海拔 2000~3700 米的针阔叶混交林和亚高山暗针叶林内，林层下生长有茂盛的竹林，这正是它们的主要食物。另外，大熊猫喜欢在相对平缓的山坡上活动，可以相对减少能量的消耗；它们的生活区域附近往往有溪流可以饮水。由于大熊猫是恒温动物，它生活的海拔高度随着季节的变化亦有差异。

130. 大熊猫分布区的气候如何？

答：大熊猫分布区气候为典型的山地温暖潮湿的气候，夏无酷暑，雨量充沛。

131. 大熊猫的领域面积有多大？

答：在野外，不同性别、不同年龄的大熊猫，其领域面积受人为因素、森林覆盖度和可食竹数量等的影响不同，大小不一样。雄性大熊猫的领域一般比雌性的稍大，约为 6~7 平方千米；亚成体的领域一般为 4~6 平方千米。

132. 大熊猫栖息地大面积竹子开花一般会多少年发生一次？

答：竹子开花是指竹子在每 30~120 年大面积开花而进行的有性繁殖现象。大熊猫栖息地某一区域的一种或几种竹子大面积开花后枯萎死亡，接下来经过数年的种子恢复，逐渐再萌发生长。竹子开花死亡后，通常要经过近 10 年的周期才得以恢复竹林，在这一恢复时期，大熊猫必须找到其他没有开花的竹子取食。

133. 大熊猫栖息地大面积竹子开花会威胁大熊猫种群的生存吗?

答: 由于大熊猫主要依赖竹子为生, 因此, 竹子生长周期给大熊猫带来一定的生存问题。当一种竹子大面积开花时, 就会增加大熊猫获得食物的难度。如 20 世纪 80 年代中期, 大面积的竹子开花造成了许多大熊猫的死亡。但是, 在大熊猫生活的栖息地里, 通常至少有两种以上的竹子。当一种竹子由于开花而食物供应减少时, 大熊猫可以迁移取食另外一种竹子, 或者扩大自己的领地, 取食没有开花的竹子。但是, 由于大熊猫的栖息地被人类活动的干扰而被破碎化, 大熊猫在不同栖息地之间的迁移受到阻隔, 因此, 大面积的竹子开花对现在的大熊猫种群生存来说, 仍是一个潜在的威胁。

History of Giant Pandas

大熊猫的历史故事

GIANT
PANDA!

134. 西方世界是从什么时候开始认识大熊猫的？

答：1869 年 3 月 11 日。

135. 第一个发现大熊猫的西方人是谁？他是从事什么职业的？

答：他是法国的阿曼德·戴维（Armand David）；他既是传教士，又是博物馆的标本采集员。

136. 第一次把活体大熊猫带出中国的人是谁？她是动物学家吗？

答：她的名字叫露丝·哈克尼斯 （Ruth Harkness）；她既不是动物学家，也不是动物园的技术人员，而是一名美国的服装设计师。

137. 露丝把捕到的熊猫幼仔取名"苏琳"？

答：露丝当时误认为这只大熊猫幼仔是雌性，故以一位随行的中国人的夫人名字为其取名。

138. 第一只活体大熊猫到美国后活了多久？

答：它在美国仅活了一年多时间，于 1938 年 4 月死去。

139. 大熊猫是如何被发现的？

答：说到大熊猫的"发现"，就会提到法国传教士戴维，因为是他在 1869 年发现了这一让当时西方世界都难以认同其存在的"奇怪"动物，很多人认为世界上不可能有这样"构思"特别的生灵。庞大的身躯像头熊，但黑白分明；圆圆的白脑袋，又镶着一对黑眼窝和一对黑耳朵；黑色的前后肢被白色的背腹泾渭分明地隔开，宽扁的尾巴紧包臀部，像是无尾。其实对于戴维神父的"发现"，更准确的说法应该是由他将大熊猫这一神奇物种介绍到了西方，并由确信它存在的科学家在 1870 年按现代分类学的规定给予了科学命名。其实中国人早在 3000 多年前的西周就已知道大熊猫的存在了，只是当时不叫大熊猫，而称为"貔貅"，记载其具体分布的是我国成书于 2700 年前的地理著作《山海经》："似熊，黑白兽……产于邛崃山严道县南。"由于古代没有系统的分类，各地的俗名众多，因此，大熊猫在中国历史上有了一长串的名称。

About Giant Panda's Conservation

大熊猫的保护工作

GIANT PANDA!

140. 中国在大熊猫就地保护方面做了哪些工作？

答：中国政府颁布了《中华人民共和国野生动物保护法》和《中华人民共和国自然保护区管理条例》等多项法律法规，实施了天然林资源保护、退耕还林还草、野生动植物保护和自然保护区建设等林业重点工程，不断完善大熊猫自然保护区体系，大熊猫栖息地条件明显改善，目前已建立大熊猫自然保护区 67 处。栖息地总面积 258 万公顷。开展了 4 次全国大熊猫调查。

141. 拯救濒危大熊猫物种的三种主要方法是什么？

答：一是就地保护，保护其栖息地，让其自然繁衍，为理想的方法。二是迁地保护，由动物园和大熊猫科研单位将大熊猫进行人工饲养繁殖，培育壮大大熊猫圈养种群。三是方法一和二的结合，在保护栖息地的同时，将圈养繁殖个体经过野化培训后放归野外，以充实壮大野外大熊猫小种群。

1963 年　67 处

142. 中国目前有多少个大熊猫自然保护区？

答：中国目前已建立大熊猫自然保护区 67 处，有效保护了 53.8% 的大熊猫栖息地和 66.8% 的野生大熊猫种群。

143. 最早建立的大熊猫自然保护区是哪几个？

答：中国于 20 世纪 60 年代发布了《关于积极保护和合理利用野生动物资源的指示》，并针对野外大熊猫的保护，于 1963 年最早专门划建了四个大熊猫自然保护区，分别是卧龙自然保护区、白水河自然保护区、王朗自然保护区和喇叭河自然保护区，覆盖面积约 900 平方千米。1970 年后，根据大熊猫分布，又建立了 8 个国家级自然保护区：即四川蜂桶寨国家级自然保护区、陕西佛坪国家级自然保护区、甘肃白水江国家级自然保护区、四川

卧龙自然保护区
白水河自然保护区
王朗自然保护区
喇叭河自然保护区

马边大风顶国家级自然保护区、四川唐家河国家级自然保护区、四川小寨子沟国家级自然保护区、四川美姑大风顶国家级自然保护区、四川九寨沟国家级自然保护区。20 世纪 80 年代又新增加了 23 处以保护大熊猫为主的自然保护区。

144. 我国最早的大熊猫科考调查活动始于什么年代？

答：1974 年，由四川南充师范学院生物系（今四川西华师范大学生命科学学院）胡锦矗教授牵头，组建了一支约 30 人的大熊猫野外调查研究队伍——四川省珍稀动物资源调查队，开始了我国第一次野外大熊猫调查活动（同时也是世界上第一次野生大熊猫数量的全面普查）。

145. 第一个大熊猫野外观察站建于何年何地？

答：第一个大熊猫野外观察站——"五一棚"大熊猫野外观察站于 1978 年在四川卧龙国家级自然保护区建立。这是世界上第一个大熊猫野外生态观察站。

146. "五一棚"的名字是怎么来的？

答：野外观察站当年使用帆布搭建而成，帐篷驻地与取水池之间是阶梯道路，总共有 51 级台阶，科学家将这个用帆布帐篷搭建而成的观察站称为"五一棚"。

147. 第一个大熊猫野外观察站海拔多高？

答：2520 米。

1978 年
2520 米

148. 大熊猫生态学研究的观察站还有哪些？

答：还有陕西省的佛坪三官庙观察站、洋县大熊猫观察站，四川省的唐家河白熊坪观察站、马边大风顶观察站、冕宁冶勒观察站等。

149. 我国第一个专门从事大熊猫保护研究机构是什么时候设立的？

答：在世界自然基金会的支持下，我国于 1980 年建立了"中国大熊猫保护研究中心"，从此开创了大熊猫的国际合作研究，在大熊猫就地保护和迁地保护研究方面均起到了重要的引领作用。

150. 中国保护大熊猫及其栖息地工程是哪年启动的？

答：1992 年。

151. 中国政府组织的四次大熊猫野外专项调查分别在什么时间?

答:

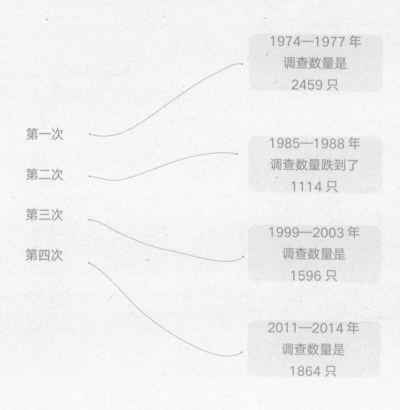

第一次

第二次

第三次

第四次

1974—1977 年调查数量是 2459 只

1985—1988 年调查数量跌到了 1114 只

1999—2003 年调查数量是 1596 只

2011—2014 年调查数量是 1864 只

152. 野外大熊猫的调查方法有哪些？

答：野外大熊猫的调查方法主要有以下几种：

咬节法：根据大熊猫粪便中的竹节长短、粗细、咀嚼程度来辨别其大概年龄、个体、种群数量、活动情况等。

DNA 鉴别法：在野外采集大熊猫粪便或毛发等组织样本，提取 DNA 做分析。现在有几种成熟的分析技术方法，可以比较准确地确定个体、年龄、性别、血缘关系等。其缺点是技术复杂，费用高。

足迹识别法：昆山杜克大学的李彬彬博士于 2017 年年底宣称建立了一种足迹模型识别法，可用于识别大熊猫个体、性别。此方法需要实践验证其有效性。

153. 大熊猫国家公园设立时间？

答：大熊猫国家公园于 2021 年 10 月正式设立，跨四川、陕西和甘肃三省，总面积 2.2 万平方千米，是野生大熊猫核心分布区。

154. 大熊猫国家公园是什么样的？

答：大熊猫国家公园位于我国第一级阶梯与第二级阶梯分界线、青藏高原东缘向四川盆地的过渡地带，地形复杂。山高谷深，水系发达。最低海拔529米，最高海拔6250米，相对高差1000米以上的深谷非常普遍。河流水系属长江流域的嘉陵江、岷江、沱江3个水系。是全球公认的地貌最复杂地区之一，同时也是地质灾害频发区。

155. 大熊猫国家公园里有什么？

答：大熊猫国家公园整合了原来分属不同部门、不同行政区域的73个自然保护地。通过加强栖息地改造修复和廊道建设等措施，增强大熊猫国家公园内13个局域种群的连通性，增加基因交流。大熊猫国家公园范围内现有1340只野生大熊猫，占整个野生大熊猫种群的71.89%。大熊猫国家公园地形地貌与生境复杂多样，分布有金钱豹、雪豹、川金丝猴、林麝、羚牛、红豆杉、珙桐等多种珍稀野生动植物，生物多样性十分丰富。

156. 为什么要抢救野生大熊猫？

答：对野外生病、饥饿和受伤的大熊猫开展救护工作，能够使野生大熊猫非正常死亡率大大降低。通过救护与恢复，将身体状况良好的被救大熊猫重新放归野外，能够有效地保证野外大熊猫种群数量。大熊猫栖息地及其周边林业主管部门、中国保护大熊猫研究中心及各大熊猫自然保护区长期肩负这项职能。至今已经有近 300 只大熊猫获救，并有"四姑娘""盛林一号"和"芦欣"等大熊猫被成功放归野外。

157. 野生大熊猫受伤或生病后是怎样被发现的？

答：野生大熊猫受伤或生病后，往往向低海拔移动，从而被巡山的保护人员或村民发现。

158. 野生病饿大熊猫抢救治愈后如何处置？

答：抢救治愈后，对于有能力回归野外生活的个体，将其放回野外；对于无能力返回野外生活的个体则将其留下人工饲养。

159. 人类为大熊猫种群复壮主要做了些什么?

答:

建立保护区, 保护其原生生态环境和栖息地。

立法对大熊猫物种进行强制性保护。

实施天然林资源保护工程和退耕还林工程, 保护扩大其栖息地面积。

建立大熊猫绿色"走廊带", 使被隔离小区的小种群能够顺利地相互移动, 形成较大的繁殖种群, 为基因交流创造条件使种群得到复壮, 阻止其遗传多样性的丧失。

迁地保护, 开展多学科的大熊猫研究工作。

进行圈养大熊猫放归野外的尝试研究。

160. 大熊猫就地保护与迁地保护工作存在哪些差异？

答：从大熊猫的整体保护事业来看，就地保护主要侧重于栖息地保护，迁地保护则以大熊猫饲养繁育与种群管理为主；就地保护的范围大、内容多、管理对象复杂，迁地保护对象单纯、研究内容相对集中；就地保护环境较差，基本在大山深处，工资和福利待遇跟不上，人才流失严重；迁地保护则离城市近，生活与学习便利，经费来源广，可以留住高学历、高层次的人才。因此，要使大熊猫保护事业得到高位发展，必须高度重视就地保护人才的引进与培养，同时也要给予野外工作者相应的待遇和荣誉的关心、爱护，充分认识到人才是推动大熊猫保护事业的主要力量。

161. 中国政府在大熊猫迁地保护方面做了哪些工作？

答：通过中外大熊猫科研工作者长期不懈的努力，在大熊猫生物学、生理学、遗传学、生态学、保护学、医学等多方面研究取得了重大进展，攻克了大熊猫繁育的世界难题。截至 2021 年年底，大熊猫圈养种群数量达 673 只；9 只人工繁育大熊猫被放归自然并成功融入野生种群。

162. 大熊猫的迁地保护能起到什么作用呢？

答：大熊猫的迁地保护是就地保护的重要补充。世界自然保护联盟（IUCN）主张，当一个动物种群在整个自然环境中的总数量下降到 1000 只（头）左右时，就有必要将一部分珍稀动物转移到适宜、安全、有保障的人工环境中。通过人工饲养、繁育，使该动物的人工种群达到自我繁衍与维系，在达到一定数量后，再进行野化放归，有计划地、科学地重建和复壮野外种群。

另外，在大熊猫的迁地保护过程中，可以方便地对大熊猫进行多学科的科学研究，了解其行为、习性、生理、疾病等方面的数据。还可以开展丰富多彩的公众教育活动，提高人们保护大熊猫及其生态环境的意识。

就地保护　　迁地保护

163. 为什么在 20 世纪繁殖大熊猫是世界难题之一？

答：其一，在 20 世纪，能够参与繁殖的雄性大熊猫数量稀少，当时国内外的雄性大熊猫大约有 120 只，能够参与自然交配的只有 1 只。其二，雄性大熊猫发情困难，人工饲养条件下进入繁殖期的雌性大熊猫一般都能按季发情，但雄性发情困难，当时也采用人工授精作为补充手段。但雌性大熊猫一年之中大约只有三四天能受孕，而且还看不出明显的生理特征。所以在人工授精时，很难准确地掌握其排卵期。如 1989 年给 18 只雌性大熊猫进行人工授精或自然交配，甚至有的是两者同时进行，而受孕的只有 6 只。其三，幼仔很难成活。一只初生的幼仔只有普通老鼠那么大，体重约 100 克，是成年大熊猫体重的千分之一。在哺乳类动物中，除了有袋类动物以外，再没有这样母子相差悬殊的动物了。幼仔双目紧闭，看不见东西，皮肤上只有稀疏的胎毛。其发育程度大约相当于 6 个月的人类胚胎，肾、大脑及一些免疫器官和淋巴组织均未发育完全。这种机体也许适应了高山峻岭、冰天雪地、细菌稀少的环境，反而适应不了人工气候环境。大熊猫的交配率低、配种受孕率低、育幼成活率低，在圈养动物中特别突出。这个"三低"，就是当时繁殖大熊猫的难题。

164. 中国如何攻克圈养大熊猫繁育"三难"？

答：经过 30 余年的努力，中国在圈养大熊猫人工繁育方面取得了巨大成就，攻克了大熊猫繁育"三难"，实现了稳定增长。通过爱心饲养、改善营养供给、提供动物福利、外源激素诱导与行为诱导，攻克发情难；通过种公兽培育、排卵与妊娠期监测、大熊猫人工采精和保存、大熊猫人工授精，攻克配种受孕难；通过培训母兽带仔、完善人工育幼技术和人工配方奶攻克育幼成活难。

发情难

配种受孕难

育幼成活难

165. 为什么要进行人工授精？

答：人工授精是圈养条件下提高大熊猫受孕率的有效方法。科研人员通过行为和激素变化来掌握雌性大熊猫的最佳配种时间，通过采集雄性大熊猫的精液对无法自然交配或自然交配不理想的雌雄大熊猫实行人工授精。人工授精能够提高雌性大熊猫受孕概率，并为实现遗传多样性提供更多选择。

131

166. 为什么要开展人工育幼？

答：大熊猫一般一胎产 1~2 仔，也有极少数产 3 仔的情况。但是通常情况下雌性大熊猫只能哺育一只幼仔，有的甚至根本不会哺育。科研人员通过模仿大熊猫妈妈育幼环境、研制人工乳以及交换育幼等方式开展人工育幼工作，保证了大多数大熊猫幼仔的存活和正常生长发育。

167. 为什么要提倡优生优育？

答：通过严格的血缘和遗传管理、避免近亲繁殖、让年龄较大和身体素质较差的大熊猫不再参与繁殖、延长幼仔和妈妈在一起的时间以学习更多的生活技能等方式，能够有效地提高圈养大熊猫的个体素质，保持遗传多样性，提高种群活力。

168. 我国是从什么时候开始人工饲养大熊猫的？

答：大熊猫作为自然界一种现存的早期生物，在我国的饲养历史非常悠久，据《史记》记载，轩辕黄帝驱貘、虎等兽与炎帝战于泉，表明当时已将圈养大熊猫用于战争。又据司马相如在《上林赋》中记载，西汉时，

在现今陕西省咸阳市附近的皇家上林苑动物狩猎场中就圈养有较多的大熊猫，这是世界上饲养大熊猫最早的记载。在唐朝时期，曾将两只活的大熊猫和 70 张毛皮送给当时的日本天武天皇，这是国外有大熊猫的最早记录。

169. 中华人民共和国成立后人工圈养的第一只大熊猫来自何处？

答：1953 年 1 月 17 日，一只大熊猫幼体在四川省都江堰市玉堂镇白马沟被成功营救，成都动物园成为中华人民共和国成立后，第一个人工饲养大熊猫的机构，并为这只大熊猫起名为"大新"。

170. 第一只人工圈养繁殖大熊猫幼仔是哪一年出生的？

答：1963 年 9 月 9 日，大熊猫"明明"在北京动物园出生，成为圈养状态下繁殖的第一只大熊猫。

171. 世界上第一对人工圈养大熊猫双胞胎抚育成活是什么时候？

答：1990年8月24日，成都动物园"庆庆"首次成功繁殖首例大熊猫双胞胎，取得了大熊猫繁育史上的重大突破，因为出生在北京亚运会前夕，所以取名为"娅娅""祥祥"，备受国内外关注。这对大熊猫双胞胎成功抚育成活是当时科研技术积累的结果，从此开创了大熊猫人工圈养繁育的新纪元。

172. 圈养大熊猫的食物营养组成有哪些？

答：圈养的成年大熊猫的主要食物肯定是竹子，约占食物总量的95%。人为提供的竹子种类，大熊猫没有选择的余地，为了做到营养全面，必须投喂一些精料。以补充食物营养成分的不足。精料大多为植物性原料组成，比如小麦、大豆、玉米、大米、麦麸等。精料大概成分为蛋白质6%~17%，粗脂肪3%~4%，碳水化合物75%。另外，还要投喂一些水果和蔬菜。

173. 竹子在投喂前需要冲洗吗？

答：从储存处取回的竹子，多少会受到污染，因此用洁净的自来水冲洗一遍即可投喂，不需要消毒。

174. 每日给大熊猫投喂竹子的总量是如何控制的？

答：每日投喂竹子的总量要根据大熊猫的大小和食量而定，经测算，大熊猫食竹量大概为其体重的 6% ~15%，一般每只成年大熊猫每天需要准备 2~3 次投喂，包括夜间。

175. 圈养大熊猫的活动场有哪些具体要求？

答：活动场面积一般要求达 300~500 平方米，活动场四周隔离墙的绝对高度不低于 2.8 米，周边墙体光滑，不能有缝隙和突出点。如果考虑游客参观，可以下挖 1.2 米，上部用 1.2 米的光滑玻璃隔离大熊猫。在沟壑和隔离墙周边布设防逃匿电网，设置脉冲电压为 700~1000 伏或 5000~10000 伏，保持 24 小时通电，形成安全双保险。在白天或有人时，可以设为低压脉冲模式；夜间或警戒级别高时，可以调整为高压脉冲模式。

大熊猫活动场地的富化、丰容也是不能忽视的重要环节，活动场内最好有起伏，应种植 5~10 株冠幅较大的乔木，再配置一些灌木，地面用适合当地气候条件的草坪覆盖。周边最好不砌高墙，避免影响空气流动。另外，要考虑给排水，活动场中央区域可用石材修建一个 4~5 平方米的水池，水从高处流下，水池可供大熊猫饮水和洗澡。活动场内排水要通畅，明暗沟结合，要考虑竹子残渣过滤网的设置，避免排水系统堵塞。

176. 圈养大熊猫环境富化是怎么一回事？

答：把一个相对自由的动物隔离在单一、有限的环境中饲养，对濒危动物来说，虽然是找到了一个避难所，但从动物习性来看，它会失去很多的自然属性和自由。这就给动物饲养管理机构提出了一个要求，在圈养大熊猫的时候，要尽量富化环境，即按照大熊猫的生活习性，在兽舍和活动场所科学地设置一些物件和栽种一些植物，既能使大熊猫可以躲避强烈的光照，又能沐浴温和的阳光；既能在地面漫步，也能攀爬树木；既有开阔的奔跑空间，也有隐蔽的场所，可以避免一些人类的刺激与干扰。

177. 怎么识别圈养大熊猫的个体？

答：大熊猫饲养机构的饲养人员都能够识别自己长年照顾的大熊猫，他们主要依据大熊猫的年龄、体格大小、面部特征、被毛深浅等特征来区别不同的个体。另外，在大部分大熊猫的口腔或皮肤上都做了刺青编号标记，还要在大熊猫皮下埋植电子标签，通过专门的电子读卡器就可以读取大熊猫的档案资料。据最新消息披露，研究者还成功开发出了熊猫脸谱识别技术系统，以后对大熊猫的个体识别将会更加简单、方便。

178. 大熊猫无线电项圈有何作用？

答：第一，通过三个以上定位点监测后，就可以准确地知道大熊猫在什么地方，并知道它是否在活动，用这种方法可以摸清大熊猫的活动规律。众多信息表明，大熊猫一般是单独活动，不固定栖息在一个地方，但活动范围很小，经常是不分昼夜不停地取食、游荡和短时间的休息。第二，在不同季节，通过无线电讯号脉冲频率的异常变化，可以了解到大熊猫的发情、交配、繁殖等情况。第三，可以进一步研究大熊猫的种群结构。

179. 人工饲养繁殖的大熊猫放归野外的难点是什么?

答: 难点主要是教会大熊猫如何在野外建立自己的领地、如何选择寻觅食物来满足自身营养的需要、如何认识并逃避天敌和如何应对体外寄生虫等。

180. 为什么要把大熊猫放归野外?

答: 将人工繁殖的大熊猫通过野化培训后放归野外,使其融入野外大熊猫种群,改善局部地区野生大熊猫的遗传多样性,增加野生大熊猫的数量,提高野生大熊猫的生存能力,是开展大熊猫人工饲养、繁育和科研的重要目的。

181. 人工饲养大熊猫野化放归的第一步工作是什么?

答: 人工饲养的大熊猫,是在人类精心呵护下成长起来的,它对人的依赖性比较强。在人工环境条件下,首先要保证大熊猫的健康,然后尽可能地想办法繁殖,在较短的时间内尽快增加大熊猫的数量。在追求数量的同时,如果忽略了大熊猫综合质量的提高,大熊猫在人为干预和受限的环境中，会丢失很多天然行为。恢复大熊猫丢失的行为是一件相当不易的工作，相对来说，丢失容易，恢复起来却非常困难。创造条件恢复大熊猫丢失的行为就成了野化放归环节中的重要起始点。

182. 达到什么目标才能算大熊猫野化放归成功？

答：经过长期野化训练，使人工圈养大熊猫恢复野性，将它放归到大自然中，大熊猫能够适应野外的环境，并能正常健康的生活，就算达到了野化放归的第一步目标；在自然界中，野化放归的大熊猫能与周边伙伴及伴生动物和谐相处，并能够规避个别掠食动物的威胁和寄生虫、疾病的侵袭，就算是达到了一个更高的层次；不仅如此，在自然界通过竞争，能够找到自己的中意伴侣，并顺利进行交配，成功产子且抚育成活，进入持续繁衍的状态，这才是野化放归的最终要求和目标。

183. 首次圈养大熊猫野外放归培训是在哪一年？在什么地方进行的？

答：2003 年 7 月 8 日，在卧龙自然保护区核桃坪。

184. 大熊猫"祥祥"是什么时候被正式放归野外的?

答:2006年4月28日。

185. 首只母兽带仔野化培训后放归的大熊猫叫什么名字?

答:这只大熊猫,就是"淘淘"。

186. 放归自然的大熊猫有多少只？它们在野外生活得怎么样？

答：中国从 2006 年开始实施大熊猫放归自然活动。到 2018 年底实施人工繁育大熊猫放归自然活动 8 次，放归野外 11 只。

2006 年 4 月放归的"祥祥"

2012 年 10 月放归的"淘淘"

2013 年 11 月放归的"张想"

2014 年 10 月放归的"雪雪"

2015 年 11 月放归的"华娇"

2016 年 10 月放归的"华妍"和"张梦"

2017 年 11 月放归的"八喜"和"映雪"

2018 年 12 月放归的"琴心"和"小核桃"

其中"祥祥"和"雪雪"未能成活，通过采集的监测数据，专家判定其他 9 只实现了圈养大熊猫自然栖息地生存的阶段性目标。

187. 大熊猫放归野外应注意哪些疾病？

答：放归大熊猫对野生种群的最大威胁，就是把疾病传播到野外，尤其是传染病，可能会给野生种群造成毁灭性的灾害。所以，在放归前应明确放归个体是否健康，必须为此进行一次全面的身体检查，而且需要注射所有人工圈养大熊猫曾经患过的传染性疾病的疫苗。在动物园或繁殖饲养中心内，往往同时饲养着其他种类的野生动物，它们携带的各种细菌、病毒都可能在动物中传播。生活在动物园的动物已经适应了这种环境，并对这些细菌和病毒具有了免疫力和抵抗能力，即便发生了疾病，也可以得到及时的救治。但是，生活在野外状态下的动物可能从未接触过这些细菌或病毒，因此，它们不可能具有相应的免疫能力。

188. 大熊猫野化放归要达到什么目标？

答：大熊猫野化放归的目标是增加小种群野生大熊猫的数量，改善其遗传多样性，消除大熊猫小种群灭绝的风险；在大熊猫历史分布区重建大熊猫种群；在实现大熊猫野生种群长期续存的同时，为其他大型兽类的保护性放归提供借鉴，保持并恢复自然生物多样性，促进当地和国家长期的经济社会发展，促进和提高全民保护意识。大熊猫放归野外后，还要对其进行持续的科学检测和研究，包括对栖息地利用模式、活动节律、觅食行为、肠道微生物菌群、激素变化和疾病的研究等。

189. 大熊猫野化放归可以借鉴其他物种野化放归的经验吗？

答：目前，国外已开展了大量有关大熊猫近源种如美洲黑熊、棕熊等的野外放归研究，这为人工圈养大熊猫放归工作的开展提供了可资借鉴的经验。

190. 我国目前有哪些大熊猫野化放归研究基地？

答：目前有中国大熊猫保护研究中心核桃坪野化培训基地、中国大熊猫保护研究中心天台山野化培训基地、成都大熊猫繁育研究基地、都江堰繁育野放研究中心和楼观台大熊猫野化培训基地，以及栗子坪大熊猫放归基地、大相岭大熊猫生态适应性放归基地两个野外放归基地。

191. 现在已具备大熊猫野化放归的条件了吗？

答：综合分析，人工圈养大熊猫放归野外的条件已基本具备。大熊猫放归前期工作早已启动，2005 年、2006 年先后将大熊猫"盛林 1 号""祥祥"放归，前者是抢救的野外大熊猫，异地放归后，已经有了生活领地。之后的人工圈养大熊猫"祥祥"，虽然在放归不到一年后死亡，但为人工圈养大熊猫的野外放归探索了道路，这是大熊猫成功放归必须付出的代价。2012 年 10 月 11 日放归的"淘淘"，5 年之后回捕体检证明健康良好，说明母带子野化训练是非常有效的。

192. 世界上有多少国家正在开展大熊猫合作？

答：中国现与美国、日本、奥地利、泰国、英国、法国、比利时、西班牙、芬兰、德国、荷兰、丹麦、新加坡、马来西亚、印度尼西亚、韩国、澳大利亚、俄罗斯、卡塔尔等 19 个国家 23 个合作单位开展了大熊猫国际合作研究项目，在外大熊猫 71 只，满足了国外民众观赏中国"国宝"的愿望，提高了公众保护意识、提升了濒危物种保护能力、促进了国际间人文交流，为推进生物多样性保护发挥了积极的作用。

193. 全球大熊猫保护研究合作项目主要开展了哪些工作？

答：近些年来，我国积极推动大熊猫保护的全球行动，大熊猫科研成果实现全球共享。一方面，积极开展大熊猫保护的国际合作研究。目前已有 19 个国家、23 个动物园与我国开展了大熊猫保护合作研究项目。截至 2022 年 10 月在国外参与国际合作研究项目的大熊猫数量共有 71 只，在国外繁育存活的大熊猫幼仔有 65 只，其中 37 只已按规定回到国内。另一方面，积极搭建全球大熊猫保护合作交流平台。通过定期举办大熊猫保护国际国内会议，如大熊猫国际会议，海峡两岸暨香港、澳门大熊猫保育教育研讨会，大熊猫繁殖技术年会等，交流和探讨大熊猫保护的新方法、新技术和新成果。此外，我国还积极打造国内科研合作平台，并建立了大熊猫国家公园珍稀动物保护生物学国家林业和草原局重点实验室、四川省濒危野生动物保护生物学重点实验室。

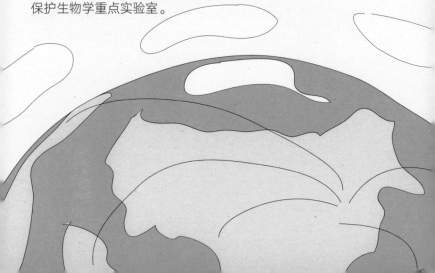

194. 国外第一次成功繁殖大熊猫是在哪个国家?

答: 1979年, 日本东京上野动物园大熊猫首次产子。

195. 采取科学研究的合作形式在哪些国家取得了大熊猫成功繁育?

答: 采取科学研究的合作形式, 最为成功的范例应该是成都大熊猫繁育研究基地与日本和歌山白浜野生动物园的合作项目。自1994年开始至2021年, 共成功繁殖大熊猫17只, 这是迄今为止大熊猫在国外繁殖最多的。美国的圣地亚哥动物园、华盛顿动物园、亚特兰大动物园, 西班牙的马德里动物园, 法国的博瓦勒动物园, 奥地利的维也纳美泉宫动物园, 日本的上野动物园, 泰国的清迈动物园都成功繁殖过大熊猫。

196. 第一只在国外出生又回到中国的大熊猫叫什么名字？

答：第一只在国外出生的大熊猫"华美"于 1999 年 8 月 21 生于美国圣地亚哥动物园，4 岁半时回到四川卧龙保护区。它的父母叫"白云"和"高高"，是 1996 年赴美参加中美大熊猫合作研究项目的。"华美"回国后先后生了三次双胞胎，因此称它为"英雄母亲"。

197. IUCN 宣布将大熊猫的濒危等级从"濒危"降为"易危"，保护力度会降低吗？

答：2016 年 9 月，IUCN 在第六届世界自然保护大会上宣布将大熊猫的濒危等级从"濒危"降为"易危"，引起学界、管理部门及舆论媒体等方面的高度关注，大熊猫被"降级"反映了中国政府对大熊猫保护所做出的努力和取得的积极成效，体现了国际社会对中国大熊猫保护成绩的认可。尽管大熊猫保护取得了积极的成效，但栖息地破碎化导致种群隔离依然是大熊猫保护面临的重大挑战，栖息地保护与恢复、大熊猫圈养种群遗传多样性、大熊猫疾病防控水平和能力、大熊猫保护管理能力都有待进一步加强和提高，大熊猫所受的威胁以及濒危状况仍然不可忽视。中国政府主管部门强调大熊猫保护级别不会降低，保护力度不会减弱，仍是中国濒危物种保护的旗舰种和伞护种，将

继续坚持不懈地按照国家一级保护野生动物和《濒危野生动植物种国际贸易公约》附录I物种的保护要求，加强大熊猫保护工作。

198. 大熊猫会从地球上消失吗？

答：从目前情况看，中国政府高度重视大熊猫保护工作，不仅重视大熊猫的就地和迁地保护，还建立了大熊猫国家公园，为大熊猫栖息地的完整性和原真性的有效保护打下了坚实基础。所以大熊猫将与我们人类同在，我们的子孙后代都有机会看到大熊猫。

199. 保护大熊猫的现实生物学意义在哪里？

答：大熊猫在《中国国家重点保护野生动物名录》里是一级保护野生动物，是明显的"伞护动物"，从一定意义上讲，保护好大熊猫及其栖息地，就是保护岷山、邛崃山、凉山、大相岭、小相岭及秦岭这六大山系的所有野生动物、生态环境和生物多样性。

从大熊猫的科研价值、社会影响力以及对生态环境保护的推动作用来看，做好以大熊猫为首要物种的野生动物保护事业，将有利于促进大熊猫及其栖息地的环境改善和植物群落的完整性，进一步拓展生物多样性的丰富度。尤其是持续不断地建设与打通大熊猫栖息地走廊带，极大地推动了不同山系和相对孤立的野生大熊猫种群之间遗传基因的交流，对维护种群健康和数量稳定增长起到了决定性作用。

200. 保护大熊猫的未来生物学意义是什么？

答：虽然对大熊猫的生物学研究取得了不少成果，但作为一个物种，人们对它的了解还远远不够，严格意义上讲，只是一知半解，空白点很多。到底大熊猫这个物种身上携带了多少生物学信息？现在还是未知数，比如它的演化过程、适应过程以及生存对策，还有跟它一路走来的体内微生物。弄清楚这些大熊猫的生存信息，对大熊猫本身的物种保护和其他动物的保护都有重大的生物学意义。大熊猫会对将来的人类社会与人类健康产生什么影响，也是不可预见和不可估量的。比如在大熊猫幼仔成长过程中起着关键性作用的母乳，其中含有非常重要的免疫因子，以及很多尚不清楚功能的蛋白质成分。或许，这些成分将在未来人类健康、延寿或美容方面起到震惊世界的作用。

201. 为什么说保护大熊猫就是保护人类自己？

答：野生大熊猫栖息地是全世界 34 个生物多样性热点地区之一，大熊猫又是分布区内的顶级物种，人们也叫它伞物种。大熊猫栖息地同样也是很多其他动物的栖息地，保护大熊猫也就同时保护了同地域的其他若干物种，使得同一地域的整个生态系统都得到保护与恢复。保护大熊猫的同时也保护了其他生物，也就保护了整个生物多样性和生态系统。而良好的生态系统能为我们不断提供青山绿水，新鲜空气，鸟语花香等公共产品，满足人类的最基本福祉。因此，人们常说保护大熊猫就是保护我们人类自己。

GIANT
PANDA! 大熊猫 201 问

Top 201 Questions About Giant Panda

Throughout a splendidly complicated evolution history of the earth, there have been hundreds of millions of species. Many species have been eliminated by natural laws over a long period of time. Some, however, have gradually adapted to nature through evolution and have continued to this day. Giant panda is one of them. From its early species called *Ailurarctos lufengensis*, the body size of the giant panda has evolved from small to large and then small again, spanning a period of about 8 million years.

65 million years ago, dinosaurs went extinct

8 million years ago, the early species (*Ailurarctos lufengensis*) of giant panda appeared

2 million years ago, *Ailuropoda microta appeared*

1 million years ago, *Ailuropoda melanoleuca baconi*, the largest panda species, appeared

12,000 years ago, the existing panda species (*Ailuropoda melanoleuca*) appeared

01 Forest of Life

The evolution of living species is like the existence of a spider web. When a species goes extinct, what left is an hollow point, and starting from this point, it will trigger the next extinction of other points ... What you can see is the extinction of a species; What you cannot see, perhaps, is the collapse of the whole ecosystem.

It is our luck that the evolutionary process of human beings and giant pandas are intertwined, so that we can live in the same world with them. Among the many species in the world, giant pandas have survived on the earth for more than 8 million years. Such a long evolutionary process has generated its unique genetic code.

At the eastern end of Hengduan Mountains, there is a long and narrow green corridor bounded between the Yangtze River and the Yellow River. It is called the Giant Panda Corridor. The Giant Panda Corridor stretches across China's Sichuan, Shaanxi and Gansu provinces. This corridor is also the only natural habitat of giant panda in the world.

Thanks to its rich and diverse biological environment, various kinds of creatures build their own homes in the Hengduan Mountains. Here, among the lofty mountains, the streams are deep, and human activities almost disappears, creating a tranquil haven for an ancient species like giant panda to live and survive.

The forest, standing at an altitude of about 2,000-3,700 meters, receives rich rainfall and is very humid.Bamboo forests are all over the place, with cloudy and foggy mist all year round. The temperature is below 20°C throughout the year. And it is this bamboo that serves as almost the only food source that can provide energy for giant pandas all year long.

02 Little Miracle

An adult giant panda weighs 90-130kg, while the average weight of a newborn giant panda cub is about 100g, which is one-thousandth of its mother's weight. It is hard to imagine that such a "super tiny little" can gradually grow into a "super big man". In fact, whether it can grow into adulthood depends entirely on the care from its mother.

With eyes closed, unable to hear or see, only a weak sense of smell, this is what a giant panda cub looks like just after being born.

A newly born giant panda cub is not black and white, but all pink skin, with only a thin layer of fluff.

After 7 days, the skin around its eyes, ears, forelimbs, and shoulders begins to darken slightly, and the white fluff begins to thicken.

After more than 40 days, the giant panda cub gradually opens its eyes. This is its first "hello" to the world.

After more than 60 days, the cub can hear.

After more than 90 days, the cub looks the same as the giant panda in our impression, and it begins to learn to walk.

After more than 120 days, the cub begins to roll, play, climb trees and learn to eat bamboo.

After about 180 days, it begins to gain weight rapidly, and as its body grows stronger, its range of activity increases.

Like every life in nature, the cub is a miracle of nature. In the spring, the newly pregnant giant panda begins to hunt for food, and she constantly eats so as to acquire the necessary nutrients. During the first 5 months of pregnancy, she will eat bamboo shoots in the early spring, and then slowly start eating bamboo leaves. During this period, in order to make sure the cub in her womb can grow healthily, she eats a tremendous volume. Meanwhile, she will also seek sites for her den at higher altitude. Usually, she will make her den inside a hole at the bottom of a big tree, or under a huge rock, and there must be sufficient water sources and bamboo forests around. Once the location of her den is determined, she will find many bamboo joints, branches and leaves, and spread them at the bottom of the den. A cozy den with bamboo leaves and moss will surely be comfortable to her cub.

During the 14 days after the cub is born, the mother panda barely eats. She will keep the cub in her arms until it reaches one month old. If the cub feels hungry, thirsty, cold or hot, or if its mother is holding a little too tight, it will make different sounds to remind mother to take good care of cub.

03 Scenes from Childhood

For giant panda cubs, childhood is the most important and happiest time. During the year they spend with their mothers, they not only have to acquire various survival skills, but also accumulate some knowledge about the living environment around them. They will travel through many parts of the forest, sometimes sprinting on the hillsides, sometimes frolicking in the streams.

When a cub reaches about three months old, the mother will take it to the entrance of the den. At the beginning, the cub could not walk, and the mother would take her newborn cub for a walking lesson by the scruff of its neck.

When the cub is able to walk on its own, it will be quite curious about the surrounding environment. At this time, the mother begins to teach it the first lesson, tree climbing and food hunting. Sometimes it lingers on the hillside, plays a little, and then looks for mother by tracking her smell down the road. Sometimes, the mother will observe its action nearby. If the cub gets lost or

the sign of her smell is lost, the cub will make a screeching cry, waiting for the mother to come back to it.

The branches are good places to avoid danger, teaches the mother panda. If the cub encounters danger, it can climb up a tree, hold its trunk firmly with the front paws, and climb hard with the hind paws.

Sometimes, the mother travels around with her cub. It needs to inspect the surrounding environment, and to collect information about where to find delicious food and clean water source.

The mother also teaches the cub to distinguish which bamboos are delicious and which ones are bad.

The giant panda is good at telling which bamboo tastes the best in each season, and which part of the bamboo is the most nutritious and palatable.

In summer, giant panda mothers will take their cubs to high altitudes to escape the heat wave; in winter, they will not hibernate; and their thick fur help them keep out the cold in snow when hunting for food.

When they were young, giant pandas are very naughty and lively. They enjoy sitting and sliding down the slope or on the snow. Some juvenile giant pandas would even go down the mountain and enter villages in winter, taking the villagers' buckets and utensils out as toys, and throwing them away after playing to their heart's content. Anything can be regarded as a toy in their eyes. After a full meal, they either play or take some sleep.

As the time goes by, the functions of various organs of the cub gradually improve. The black eye circles around its eyes have grown to 5cm, as if it is wearing a pair of sunglasses. It also has a pair of black ears that are large and round. Its eyes are small, and its vision is slightly worse than its hearing ability. The black ear is an adaptation to the cold environment. Such color can absorb heat energy, speed up blood circulation at the ear tip, and reduce heat loss. Giant pandas are more afraid of heat than cold, and they like to live in cool places.

During this year of its growth, the cub gradually increased its food intake and spent more time foraging each day. The short but substantial exercise in this year has given it a good opportunity to grow quickly. It is becoming stronger and bigger, and it's time to go to see the world.

04 Home of My Own

For all living species, becoming a sub-adult means it's time to leave your mom and start an independent life. An inner force begins to take shape, and that is the tendency to leave its mother. Like a migratory bird, when the time comes, it must set off and fly away.

A giant panda often feels hard to leave its childhood home. However, in order to avoid inbreeding and ensure the continuity of their reproduction, its mother has to drive her daughter out but keep her son nearby.

Each giant panda will have a home of 4 to 7 square kilometers. This is the area he chooses after having inspected many places. There needs to be lush bamboo forests, abundant food, streams and clear mountain springs around its home, and it should also be very quiet and secluded. Although it is a small world, it can sleep tight, find its love, and raise children. Building a home is very important for giant pandas.

The vast living space makes it not so difficult for giant pandas to find such a suitable area. The giant panda will leave its smell on the boundary of its home, which can avoid conflicts with other giant pandas.

When a cub reaches two and a half years old, it will feel alienated

by its mother. Sometimes the mother will even drive it away from home. Such behavior means that it should leave the area and find and establish its own home.

Even if it is forced to leave this area, it will continue to come back here, and even if it builds a new home, it will not be too far from here. No matter where it goes, the grown-up cub can still find its former home by its smell. Although this parting process seems cruel, it is also one of the reasons why giant pandas can survive for such a long time.

The fact that giant panda using a fixed area as its home reflects a kind of survival wisdom that prevails in its long-term evolution. Each giant panda guards one side of the land, which can not only adjust the density of the population, but also avoid food competition among giant pandas.

Such wisdom can only be gradually understood and learned by giant panda cubs after they have their own homes and families.

05 Rangers in Bamboo Forest

Everything in the world deserves our appreciation and respect, but the significance of the giant panda seems to have its own unique value. Its appearance is simple, honest and cute, and it is adorable; its character is mild and attracts attention; and in its long history of evolution, it has gradually evolved from the habit of "meat eating" to bamboo eating, avoiding causing damage to human or other animals when competing for food. This fully reflects the superb wisdom of this ancient creature. But these are not important to giant pandas. Living freely in the wild is their favorite thing.

They eat day and night, they don't choose a specific place to sleep, and they don't have a fixed place for excretion. They love to roam and do not have a fixed foraging place or resting place. They roam and forage in the bamboo sea and like to live freely.

For most of the time, giant pandas live in dense mountains and forests, and they are like solitary wanderers and are hard to find.

An adult giant panda can weigh up to 90-130kg and stand about 1.7 meters tall. It looks huge and moves slowly. However, giant pandas are actually very flexible in their actions, whether they are foraging for food or climbing trees, they are masters. A giant panda can run at a speed of more than 40 kilometers per hour in the forest.

Born to climb

For giant pandas, big trees are not the only place to avoid danger, but also a place to rest, relax and play. They are well versed in climbing trees since childhood, and their joints are very flexible and can complete difficult movements like contortionists. When climbing a tree, their front limbs are attached to the trunk of the tree and slowly climb up until the hind limbs are fully stretched, and then climb onto the big tree. When descending from a tree, they keep their heads up, slowly back down, and then jump down agilely when close to the ground.

On the tree, they can breathe fresher air and enjoy the warm sunshine. They can also stand high on the tree and see what is happening below. Sometimes they stay in trees all day or even days, sleeping or sitting. When encountering their predators or chasing and playing with each other, they can also quickly climb onto high trees, and outrun their opponents.

The life on the tree is not affected by the size of their body. They can move freely on the tall, dazzling treetops, swaying, but not feeling afraid. Whenever it seems that they are about to fall from the tree, they deftly move on to the next one. When they want to sleep on the tree, they will pick up a comfortable branch, ride on it and sleep with their paws cuddling the tree, which is very comfortable and enjoyable.

A natural water lover

There is an old saying in China that the benevolent man enjoys the mountains, the wise man enjoys the water. And the giant panda living in the mountains and forests is like a retired wise

man. Giant pandas can often be seen by the creeks in the dense forest or by the babbling springs.

For a giant panda to establish a living area, it must locate a water source within 800 meters. In order to find clean and tasty drinking water, it often travels long distances from high mountains to river valleys. Once it locates the water source, it will drink abundantly. Sometimes it will lie down by the water regardless of his posture, and drink enough in one breath. Then, it will start to play and play with its companions and have a lot of fun. In summer, it will take a bath in the water, and in winter, this place is nice for some cold swim.

Giant pandas can traverse in mountains and waters, cross turbulent rivers, and sprint into the distance upon danger. They are cumbersome yet flexible, simple yet clever, and they are the kind of animal that cannot be judged by their appearance.

The vegetarian way

Wild giant pandas can always enjoy a life that is guided by wisdom and delight. Hundreds of thousands of years ago, in order to get rid of competition with other carnivores, they chose a vegetarian path and began to eat bamboo as their main diet. And the price to pay is that they spend most of their lives eating.

A typical daily schedule for a giant panda:
0:00-2:00 break

2:00-7:00 eating (breakfast)

7:00-10:30 break

10:30-11:00 roaming and playing

11:00-12:00 eating (lunch)

12:00-14:30 break

14:30-21:00 eating (dinner)

21:00-2:00 break (night sleep)

Although the giant panda, which was originally carnivores, changed its eating habits, it did not change the absorption capacity of its stomach. The nutrition contained in bamboo-based food is relatively low, and the absorption rate of panda's digestive system is also low. Therefore, they can only change their living habits and continuously eat to meet the physiological needs.

Interestingly, even if giant pandas no longer eat meat, they care greatly about the taste of their food. When they eat bamboo, they will choose carefully, just like masters of bamboo tasting. No matter the season or geographical region, bamboo shoots are always their favorite. The shoots are very rich in water and nutrients, with the highest sugar content, making them sweet and very delicious to eat.

In spring, when new bamboo shoots emerge, giant pandas will enter the bamboo forest, feeding on them. In summer, they will turn to bamboo shoots that are one or two years old. In autumn, bamboo leaves become their favorite. And in winter, they tend to eat old bamboo shoots. Giant pandas are not only picky about the freshness of bamboo, but also care about location and species of bamboo. The bamboo species they like is located at an altitude of 1800-3200 meters.

After choosing their direction and location for feeding, the giant pandas will check the bamboo area by area, and then branch by branch. Typically, they will crawl in a zigzag or "8"-shaped route pattern, so as not to miss the best bamboo. When the food is ready, the panda finds an open space, sitting upright or lying down to eat.

Although they look slow, they are quite efficient when it comes to eating. There's no extra movement in choosing, preparing and eating the food. The whole process is full of efficiency. They bite quickly, chew only if necessary, and often stretch out their front claws to grab the bamboo before the previous bite is swallowed. Their wrist bone, after a long period of adaptation and evolution,

has also evolved into a finger that can hold the bamboo firmly.

The evolution of giant pandas is a complex process, with evolving habits leaving marks on their ever-distinctive characteristics. They have been making changes that are beneficial to their survival in the changing environment. The only thing that remains unchanged is their souls which are always free in the wild forests for eight million years.

06 Neighbor of Human

The best way to save a biological species is to preserve the integrity and stability of its population, as well as the genetic diversity within the population. Today, in terms of the protection of giant pandas, the most fundamental work we must finish is to preserve wildness and freedom of their populations.

The cuddly giant panda has evolved from stages like *Ailuropoda microta*, *Ailuropoda milanoleuca daconi* and finally takes on its modern look. It once bravely faced the sea and climbed mountains after mountains. It was once accompanied by the existence of the legendary saber-toothed tigers and saber-toothed elephants. And now, they become human being's companions all the way through human evolution. It has become one of the most famous "living fossils" of the animal kingdom in the history of biological evolution.

Today, the giant panda is not only a rare species in China, its protection work has also become one of the demonstrations of the modern conservation of endangered species. The symbol of the giant panda has long been integrated with Chinese civilization and culture, and has become China's "national treasures".

In the 1980s, many zoologists in China and around the world put the protection of giant pandas at the forefront. At the same time, many field researchers who have devoted themselves to the wilderness for decades, have also emerged to follow the trail of giant pandas. The places where giant pandas live are often in the deep mountains. They are very good at covering up their tracks. Therefore, the protection work of wild giant pandas is extremely difficult to carry out.

For China, giant pandas have both ecological and cultural values. For most of the workers who have devoted themselves to the protection of giant pandas whether these pandas can continue to reproduce and whether this population can continue to grow is what they really care about. Some of them have spent most of their lives, and some have even given their lives, but they still spare no effort to work for the conservation project for wild giant pandas.

To this day, the international community's knowledge of giant pandas has reached an unprecedented level. From 1999 to 2003, China completed its third national general survey of giant pandas. In February 2015, the National Forestry Administration of the People's Republic of China announced the results of the fourth national general survey of giant pandas. The result showed that compared with 1596 pandas of the third survey, China's wild giant panda population reached 1,864. The population is growing steadily with a reasonable biological structure.

Meanwhile, we also have to see that the giant panda still faces many threats such as its overall population being too small, human disturbance, habitat fragmentation, diseases and natural predators. While maintaining the momentum of the protection and research work on their population and habitat, *ex situ* conservation has become a necessary supplement to *in situ* conservation. Scientists have successfully overcome the breeding challenges of captive giant pandas through unremitting efforts. At present, the population of captive giant pandas has reached more than 600. The genetic diversity of this population also continues to increase. At the same time, for *ex situ* conservation workers, releasing the captive giant pandas into the wild has been a long-held dream. Animals belong to nature, and the giant pandas that live freely in the mountains and forests

are the real giant pandas. Since 2003, China has launched the "wild reintroduction" program of giant pandas. Up to now, 9 captive-bred giant pandas have been released into the wild and successfully welcomed by the wild population.

We and giant pandas live by the same river and drink the same water. Therefore, these conservation projects not only protect their habitats but also protect ourselves. The most spectacular and precious wildlife communities in the world are being formed in China's mountainous areas like the Qinling Mountains, Minshan Mountains, Qionglai Mountains, Xiangling Mountains, and Liangshan Mountains. Animals such as giant pandas, takins, golden monkeys and other animals reproduce in these habitats and continue to thrive, making these lands full of vitality. From field tracking to habitat protection monitoring, from *ex situ* protection to wild release, our conservation effort has evolved from accumulating knowledge of giant pandas to actual panda breeding, and ultimately, we will make them a sustainable species in a complete, continuous and broader homeland.

In 2021, the opening of China's Giant Panda National Park further enhanced the connectivity and integrity of the giant panda habitats, providing a better ecosystem for the survival and reproduction of giant pandas. This Park will also contribute a new model to the international community, showcasing China's ecological conservation effort.

Meet the
Giant Panda

Q: Who are the direct ancestors of giant pandas?

A: *Ailurarctos lufengensis*.

Q: How old is the giant panda species?

A: Judging from the current fossil record, *Ailurarctos lufengensis* has a history of more than 8 million years.

Q: What does *Ailuropoda melanoleuca baconi*, look like?

A: Judging from the fossil records, the evolution of the giant panda's body size is from small to large, and then small again. Throughout panda's evolution, *Ailuropoda melanoleuca baconi* has the largest body size. It is 1/9 to 1/8 larger with other characteristics similar to the living giant panda .

GIANT
PANDA! 大熊猫 201 问

Q: When was the giant panda family most prosperous?

A: In the middle and late Pleistocene.

Q: What is the first book to record giant pandas?

A: Among the ancient Chinese books, the earliest records of giant pandas are found in *Shangshu* and *Book of Songs*. They are about 3,000 years old.

Q: Throughout China's history, what were the names for the giant pandas?

A: During the Yellow Emperor period, giant pandas were called *Pixiu*; during the Warring States Period, they were called *Mo* White Panther and Shi Tieshou (Iron-eating Beast); during the Three Kingdoms period, they were called *Pixiu* during the Western Jin Dynasty, they were called *Zouyu*; during the Tang Dynasty, they were called Bai Xiong (White Bear); during the Ming Dynasty, they were called *Mo*.

Q: What is the historical distribution of giant pandas?

A: During the Pleistocene period, giant pandas entered its prosperous era and were widely distributed in most parts of China. Their existence spread northward to the Zhoukoudian area of Beijing. In addition, they were also distributed in Myanmar, Vietnam, Laos, and Thailand in Southeast Asia.

Q: Why are giant pandas known as "living fossils"?

A: Judging from the excavations so far, the distribution of giant panda fossils spans 14 provinces, autonomous regions and municipalities in China, with a total of 48 sites. Its geological age belongs to the middle and late Pleistocene. Giant panda was a widely distributed species at that time. Together with other animals, it formed a distinctive animal group, which paleontologists called the "giant panda saber-toothed fauna". It shows that the giant panda was in a dominant position at that time, and the population was prosperous. Ever since then, due to the great changes in the natural environment and climatic conditions, many species in the same period became extinct one after another, being unable to adapt to these changes, and were later only found in fossils. The giant pandas have lived tenaciously after a long-term survival competition. They are the precious property left by nature to us human beings, that's why they are called "living fossils".

Q: Which order do giant pandas belong to in zoological taxonomy?

A: The debate on the taxonomic status of giant pandas has been going on for more than 130 years, and it is still going on among modern taxonomists. Some scholars believe that giant pandas belong to the raccoon family or the giant panda subfamily. Most scholars believe that giant pandas are still close to bear family, but with some differences.

Therefore, today, the most recognized classification for giant panda is: giant panda (*Ailuropoda melanoleuca*), belonging to the giant panda subfamily (Ailuropodidae), under the family of Ursidae, order of CARNIVORA.

Q: Where are the giant pandas now?

A: Giant pandas are now distributed in the deep valleys of the Qinling Mountains in Shaanxi, the Minshan Mountains in Sichuan and Gansu, and in the deep valleys of the Qionglai Mountains, Daxiangling Mountains, Xiaoxiangling Mountains and Liangshan Mountains in Sichuan.

Q: What altitude range do wild giant pandas live in?

A: The wild giant pandas' activities is mainly found in the altitude between 2,000 and 3,700 meters. Affected by human activities, the giant pandas are rare below the altitude of 1,300 meters. Some giant panda's activity reaches the altitude between 3,200-3,700 meters, with the highest record reaching 4,000 meters under special circumstances like long distance travel or migration.

Q: Why are giant pandas black and white?

A: The altitude of giant panda's habitat is usually 2,000-3,700 meters, and the climate there is relatively cold. Such color scheme reflects its adaptation to the law of survival in nature. We infer that the panda's unique fur color has a series of functions that allow it to match its background in different environments and communicate using facial expressions. The face of the giant panda is milky white, the waist and back are white; there are oval black circles around the eyes; the hair color of the ears is black; the area from the forelimbs to the part below the shoulders, and the outside of the thighs of the hind legs are black.

Q: Are there colorful giant pandas?

A: In addition to the common black and white giant panda, there are also other color variations. Brown giant pandas have been found in Foping National Reserve and Changqing National Reserve in Shaanxi, China, and white giant pandas have been found in Wolong National Natural Reserve, Sichuan, China.

Q: Is giant panda's tail black or white?

A: White.

Q: Is the fur of a giant panda soft?

A: The giant panda's coat is relatively rough, with some elasticity, and the bottom layer has some fluff. Its coat has good thermal insulation and moisture resistance, so giant pandas can live normally in the cold snow.

Q: How long does a giant panda live?

A: Research shows that the maximum lifespan of a wild giant panda is about 26 years, and the average lifespan is about 19 years. The lifespan of a captive-bred giant panda ranges from 20 to 30 years old, and the current maximum is 38 years old. A 1-year-old giant panda's is equivalent to a human's 3-4-year-old child.

Q: How are the age stages of giant pandas divided?

A: Infancy (from birth to 0.5 years old), youth (0.5 to 1.5 years old), sub-adulthood (1.5 to 5 years old) , adulthood (5 to 21 years old), and old age (21 years old and above).

Q: How are the age of giant pandas determined?

A: One way is to group the giant pandas according to their age based on the bamboo bites, chewing degree and dung size in their feces, so as to determine their approximate age. Another method is to identify the age of giant pandas through microscopic observation of tooth slices. Generally, researchers determine the age of giant pandas by looking at the color and wear degree of their teeth and the external morphology.

Q: How much does a giant panda weigh?

A: A wild giant panda usually weighs 90-110 kg, the heaviest one can reach 120-130 kg; a panda that is captive-bred usually weighs 90-125 kg, and the heaviest can reach 180 kg.

Q: How tall is a giant panda?

A: The shoulder height of a giant panda is usually 0.65-0.75 meters, and the hip height is about 0.64-0.65 meters.

Q: What is the length of a giant panda?

A: The body length of a giant panda is generally 1.2 to 1.8 meters.

Q: Are there differences in body size between female and male giant pandas?

A: Yes, there are differences, the male body size is slightly larger, the difference is about 10%-18%.

Q: What is the female-to-male ratio of giant pandas?

A: The female-to-male ratio of giant pandas is about 1.16 : 1.

Q: How long are panda's feet?

A: The foot length of a giant panda is 12-20 cm.

Q: How many toes does a giant panda have?

A: A giant panda has 5 toes on each of its front and back feet, and a pseudo-thumb made up of the radial sesamoid bone on the front foot. This "pseudo-thumb" helps them grasp the bamboo stem tightly and quickly feed on delicious bamboo.

Q: Why do panda cubs have such long tails?

A: The ancestors of giant pandas were carnivores, and their cubs from birth to pre-adulthood, more or less, carried of their ancestor's characteristic—long tails. During the growth and development of giant pandas, their tails will gradually stop growing, and the tails of giant pandas are much shorter in adulthood.

Q: Why do giant pandas have shorter tails when they grow up?

A: From birth to adulthood, the length of the giant panda's tail increases very little compared to its body, mainly because its width increases. Compared with the body length and huge size of adult giant pandas, the flat tail is not very visually prominent, which has caused many people to misunderstand that giant pandas have no tails.

Q: What is the purpose of the giant panda's short and wide tail?

A: The giant panda's tail is relatively short, about 10-12 cm in length, and it is furry and usually sticks to its body. When the tail glands, perianal glands and vulva secrete secretions, pandas can use their tails as a brush to make marks around.

Q: What is the purpose of the giant panda's frequent mark making behavior?

A: Each giant panda will mark its scent with perianal glands on the trees or stones at the edge of its territory to warn trespassers not to enter. The purpose of this marking behavior is to inform other giant pandas of its individual information as gender/age. Also, a panda can interpret from it the physiological state such as whether other giant pandas are in estrus by sniffing their scent markings.

Q: What are the changes in the appearance and organs of giant pandas from ancient times to modern times?

A: Anatomically, no matter its external shape or part of its organ structure, the giant panda has both retained ancestral characteristics and adopted adaptive changes. The most obvious is that in order to adapt to the staple food bamboo, the giant panda, from the skull, teeth and limbs to the digestive tract, has undergone some changes. The organ changes are mainly the enlargement of the chewing surface of the molars and the evolution of the sixth finger.

Q: What is the normal body temperature of a giant panda?

A: The normal body temperature of giant pandas is 36.5-37.5°C.

Q: What is the normal breathing rate of a giant panda?

A: The normal breathing rate of a giant panda is 16-24 times/min.

Q: What is the normal heartbeat of a giant panda?

A: The normal heartbeat of a giant panda is 70 beats per minute.

Q: How many pairs of chromosomes does a giant panda have?

A: A giant panda has 21 pairs of chromosomes, of which 20 are autosomes and 1 pair is sex chromosomes.

Q: What is the brain capacity of a giant panda?

A: Although the head of the giant panda looks big, but its brain is not well-developed. Its brain capacity is only 310-320 ml, which is 60 ml less than that of the wild bear.

Q: How many teeth does a giant panda have?

A: Giant pandas have 24 deciduous teeth and 40-42 permanent teeth.

Q: How many vertebrae does a giant panda have?

A: A giant panda has a total of 42 or 43 vertebrae.

Q: How many pairs of ribs does a panda have?

A: The ribs of a giant panda are asymmetrical, with 13 on the right and 14 on the left or 14 on the left and 13 on the right.

Q: How many pairs of cranial nerves does a panda have?

A: A giant panda has 12 pairs of cranial nerves. In order: olfactory nerve, optic nerve, oculomotor nerve, trochlear nerve, trigeminal nerve, abducens nerve, facial nerve, auditory nerve, glossopharyngeal nerve, vagus nerve, accessory nerve, hypoglossal nerve.

Q: How many pairs of spinal nerves does a panda have?

A: A giant panda has 34 to 35 pairs of spinal nerves.

Q: Is there a difference in the face shape of giant pandas?

A: According to sample records found among different giant pandas, their face shapes can be simply divided into round face and long face. The ears and eye rims of individual pandas vary in size. Today's giant panda fans have a strong panda recognition ability, they can identify each individual panda by its characteristics of the body parts. At present, relevant researchers have established a giant panda facial recognition system, which can identify any panda registered into the giant panda pedigree database.

Q: How do giant pandas adapt to environmental changes?

A: Giant pandas have adapted to the strategy of survival of the fittest. Their ancient body structure and organs of carnivores have undergone adaptive degradation to adapt to the objective conditions of the staple food bamboo.

Q: Are wild giant pandas vegetarians?

A: In terms of animal classification, giant pandas are classified as carnivores, but giant pandas can be said to be a typical idiosyncratic carnivore for its mainly vegetarian diet. In the wild environment, only about 1% of the giant panda's diet will include other plants and animals; in the human breeding environment, the giant panda can also eat honey, bird eggs, sweet potatoes, bush leaves, oranges, plums, bananas, etc.

Q: What is the panda's recipe like?

A: Giant pandas mainly eat bamboo and bamboo shoots, which account for about 99% of their total food. Only about 17% of them are absorbed. In the wild, giant pandas occasionally eat some other plants and animals, so the giant panda is an omnivorous animal whose main food is still bamboo. Giant pandas are very fond of water. Sometimes they even have difficulty walking due to drinking too much water. Their drinking habit also helps to their digestion of bamboo.

Q: How long do wild giant pandas spend eating bamboo every day?

A: According to research, giant pandas in the wild spend 10 to 14 hours a day eating bamboo.

Q: How much bamboo does a giant panda eat in a day?

A: Giant pandas live in cold plateau regions. In these places, the spring is usually from April to June, summer and autumn from July to October, and winter from November to March. In spring, giant pandas feed on fresh bamboo shoots, and their food intake is also the largest in this season. Generally, they feed 30-60 kg themselves, with an average of about 40 kg. In summer, their food becomes mainly bamboo stems, but also include some older bamboo shoots and newly grown shoots and leaves. Autumn is the most lush time for young branches and leaves, which makes them the main recipes for pandas. At this time, the nutrients of young shoots and young leaves rank first among other parts of bamboo, and the crude fiber contained is lower than that of bamboo stems. Therefore, in autumn, giant pandas only need to eat 10-14 kg per day, and their range of activities is the lowest in a year, which also saves the energy consumed.

Q: Why do wild giant pandas eat a lot of bamboo?

A: Although giant pandas feed on bamboo, their digestive tract is typical of carnivores. The digestive tract is short, lacks mucous membranes, and has no cecum, so it cannot effectively digest cellulose and lignin in bamboo cell walls. Their digestion can only absorb cellular contents and part of hemicellulose. Therefore, giant pandas are different from other herbivores, which have a complex ruminant stomach or a large cecum, which can fully digest and absorb the vegetables these animals feed on. That's why, in order to meet nutritional needs, giant pandas spend a lot of time every day eating bamboo in large quantities.

Q: What kind of bamboo do pandas like to eat?

A: In addition to choosing the optimal composition of seasonal recipes, giant pandas also take optimal choices in feeding behavior. In the season best for eating bamboo shoots, they will choose the best according to the location. The giant panda group in each mountain system has a favorite type of bamboo. The giant pandas in the Qionglai Mountains prefer *Bashania fangiana*. It is thinner, and giant pandas do not eat its young shoots, but only eat old shoots in winter. In the Qinling Mountains, *Bashania fargesii* is the main food choice. These bamboo shoots are thicker. Giant pandas choose to eat such bamboo shoots with a diameter of more than 12 mm. In other mountain systems, giant pandas mainly eat *Fargesia denudata*, *Fargesia robusta*, *Yushania niitakayamensis*, etc. The diameters of the bamboo shoots of these species typically exceed 9 mm.

Q: What is the typical process of a giant panda eating bamboo?

A: When a giant panda eats bamboo, it will collect bamboo leaves into a handful, eat it in bites; when eating a bamboo pole, the giant panda will use its teeth to peel off the outer skin of the bamboo and eat the inner part. During different seasons, the types and parts of bamboo that giant pandas like to eat also vary.

Q: What is the structure of the giant panda's digestive tract?

A: The digestive tract of the giant panda is the same as that of other mammals, consisting of the mouth, esophagus, stomach, small intestine and large intestine.

Q: How long is a giant panda's intestine?

A: The giant panda's intestine is divided into small intestine and large intestine. The average length of their small intestine is 490 cm (269-624 cm), and the average length of their large intestine is 120 cm (72-167 cm).

Q: How is the giant panda's intestine different from that of herbivores?

A: First, the entire intestine of giant pandas is shorter, about 4.3 times the length of their own body; while the digestive tract of herbivores is about 10 to 30 times longer than their own body length. The second is that the giant panda's colon is directly connected to the ileum, without a cecum, their colon is straight, and there is no longitudinal belt and intestinal pouch; the third is that the giant panda's intestine is rich in mucous glands, but herbivores do not.

Q: What does a giant panda's stomach look like?

A: The stomach and intestine of the giant panda is a single-chamber glandular stomach, and is in the shape of letter "U" and curved.

Q: Will hard bamboo stab the stomach and intestine of a giant panda?

A: No. Because the muscular layer of the digestive tract of giant pandas is thick, there are abundant multicellular mucous glands in the digestive tract, and the mucus secreted by these mucous glands plays a protective role in the digestive tract.

Q: How much feces do giant pandas excrete every day?

A: According to statistics, adult giant pandas defecate more than 40 times a day, and the feces excreted are about 120. If they eat bamboo shoots, they will excrete 150 groups of excrement. The weight of each group of excrement is about 100 grams, leading to a daily total of 15-20 kg.

Q: What is a typical daily schedule for a giant panda?

A: 0:00-2:00 rest; 2:00-7:00 eating (breakfast); 7: 00-10: 30 rest; 10: 30-11:00 roaming and playing; 11: 00-12: 00 feeding (lunch); 12:00-14:30 rest; 14:30-21:00 feeding (dinner); 21:00-2:00 rest (nighttime sleep).

Q: Are giant pandas only active during the day?

A: The giant pandas are not only active during the day, they can also be active at night, therefore they are both diurnal and nocturnal animal.

Q: Does light exposure have any effect on giant pandas?

A: Appropriate light exposure has a positive impact on the growth and development of giant pandas, and light can also affect the estrus time and length of estrus of giant pandas. However, strong sunlight in summer and autumn should be avoided as much as possible, especially for giant pandas after leaving their habitat. If the ambient temperature is too high and the ultraviolet rays are strong, the giant panda will suffer from heat stroke or heat related disease. Irreversible losses could be induced under such circumstances.

Q: Do giant pandas hibernate?

A: Giant pandas do not have the habit of hibernating. Since bamboo is an evergreen plant in all seasons, giant pandas can still feed on bamboo even in winter. Giant pandas usually migrate from high altitudes to mid-to-low altitude areas in winter, where the temperature is relatively high and food such as bamboo is relatively plentiful.

Q: Where do giant pandas sleep?

A: In the wild, newborn giant panda cubs stay in the warm tree hole or cave prepared by their mother; when they grow up, they like to sleep in the tree, so as to avoid the attack of their natural enemies; adult giant pandas always rest wherever they go, but they prefer to choose a place to lean on, such as by a large tree or by a fallen wood log.

Q: What sleeping positions do giant pandas have?

A: Speaking of the sleeping positions of giant pandas, there are many kinds. Giant pandas generally sleep on their backs or on their tummies from infancy to adulthood. But it is very interesting for giant pandas to sleep on tummies. Whether they are on a perch, a bed, or on a tree branch, their limbs will naturally sag, giving them a feeling of relaxation. When sleeping on the back, usually the hind limbs will be placed on a fixed object, and very often the forelimbs are used to cover their eyes, much like we use our arms to block the eyes from the light, which is very funny. When giant pandas lie down, they also stretch and yawn from time to time.

Q: What are the living habits of giant pandas?

A: Giant pandas live in dense bamboo forests for a long time and like to be alone. They usually use urine and perianal gland secretions to mark their territory. They love cool and cold. They are good at climbing trees and swimming, and do not hibernate. Once an alarming situation is detected, they will quickly hide in the bamboo forest or climb the treetops.

Q: Is a giant panda afraid of cold or heat?

A: A giant panda is afraid of extreme heat, and it is not afraid of cold.

Q: Can a giant panda swim?

A: Giant pandas are master swimmers. When the weather is hot, they can wade in the river, play in the water, and swim. They can also swim long distances in deep water, which is their main means of migrating in complex environment.

Q: Can giant pandas climb trees?

A: Yes, they are master tree climbers. When the giant panda cubs are six months old, they can follow their mothers to climb trees, live and sleep in the trees. Female giant pandas live in rock burrows or tree cavities on the ground while giving birth and caring for their cubs.

Q: Do pandas spend a lot of time lying on the tree?

A: The giant panda likes to climb after half a year old, and it is its survival instinct to like to climb trees. This is especially the case for the cubs born in the wild, when the female giant pandas have to go out to eat bamboo. During the time away from the mother's protection, the cubs will be threatened by predators, so they will quickly climb the trees to avoid them. In addition, lying on the tree is easy to get sunlight, so that the body can develop normally and healthily.

Q: Do pandas like water?

A: Giant pandas like to drink and play in water. They only prefer to drink running water in the wild, such as stream water or headwater. If the water freezes, the giant panda will break the ice with its front paws; if the water is shallow, it will dig a puddle to drink. Sometimes giant pandas drink too much water, their abdomens swelled like a drum, making them walk like drunkards. In such case, they walk or lie down by the water. That's why local people say they have the habit of "drunk by the water". Especially in summer, giant pandas will bubble their feet or sit and groom in shallow streams, and sometimes soak their bodies in water with their forelegs.

Q: What is the walking posture of giant pandas?

A: They have an intoeing walk, which means they walk with their feet turned in. Such posture has some advantages when shuttling and running in the bamboo forest. And it is also an adaptive evolution to live in the forest environment.

Q: Can giant pandas stand?

A: The giant panda is very good at standing. It will stand to feed on food in high places. When standing, it will also look around to watch, and carefully observe the surrounding movements. The most amazing thing is that male panda will also stand upside down, usually on the trunk to make the scent mark of the perianal glands. In addition, when mating with estrus female pandas, males sometimes also stand to mate.

Q: Why do pandas like to do somersaults?

A: When giant pandas are particularly happy, they like to do somersaults, rolling forward and rolling back. When doing these actions, it is in mostly sunny mornings in spring, sometimes in cool autumn days. They also play after a heavy snowfall, rolling excitedly in the snow. Sometimes he would lie on his back and spin around in circles. Giant pandas, especially their cubs, have a strong curiosity about the external environment. Therefore, they play, roll, and explore, and perceive the surrounding things by rolling and crawling.

Q: Do giant pandas also groom their coats?

A: After a giant panda plays in the water, it will shake its body from side to side to get rid of the water on its coat. Also, after eating, they will lick the periphery of the mouth and the forefoot. From time to time, it combs his ears and neck with its front paws, and also scratches with its hind legs. When it can't reach its back, it rubs against tree trunks or rocks, tickling and cleaning its coat.

Q: Why don't giant pandas like to be touched by their ears?

A: Because the ears of giant pandas are thin and their hearing is very sensitive. If their ears are touched, they will become sensitive and alert and will feel uncomfortable. If they are touched repeatedly, they will get upset and act aggressively.

Q: What is the giant panda's sense of hearing?

A: Giant pandas have very sensitive hearing, and can hear voices from far away.

Q: What is the giant panda's sense of smell?

A: Giant pandas have a very keen sense of smell, and can identify the odor information sent by other individuals.

Q: What is the giant panda's sense of sight?

A: The giant pandas, who have lived in the deep mountains and dense forests for a long time, although they can not see very far, are not born with myopia, and their visual function is far worse than his hearing and smell ability.

Q: What is the cry of giant pandas like?

A: Giant pandas make different cries under different circumstances, emotions and psychological states. Their usual cries sound like: cow barking, sheep barking, moaning, barking, chirping (like puppies), as well as howling, birdsong, squeaking and so on.

Q: Does a giant panda hurt human being?

A: Generally speaking, giant pandas are mild-mannered and do not attack other animals. But it can hurt people in self-defense. Especially in the estrus courtship and farrowing seasons, they are more aggressive and more likely to hurt people.

Q: What diseases can giant pandas get?

A: A giant panda also gets sick. The causes of its sickness can usually be categorized into several diseases, such as medical diseases, surgical diseases, obstetric diseases, infectious diseases and parasitic diseases.

Q: What are the most common diseases of wild giant pandas?

A: Parasitic diseases and gastrointestinal diseases.

Q: Do pandas also feel nervous and stressed?

A: Yes, it is typically seen in captive-bred pandas if they encounter human noise or interference nearby. If the situation continues, it will increase pandas' tension and pressure.

Q: Do pandas also get colds?

A: Cold is a disease mainly includes upper respiratory tract symptoms caused by temperature changes, decreased immunity of giant pandas and the invasion of various cold-related viruses. The observable symptoms are cough, runny nose, fever, with lethargic and sleepy feelings. General treatments include intramuscular injection of Analgin, Bupleurum, or oral Tylenol. Human colds can also infect giant pandas. Therefore, staff or researchers who suffer from colds can only continue to raise pandas after they have recovered from the disease.

Q: Do giant pandas also suffer from heat stroke?

A: The wild habitats of giant pandas are basically located in areas above 2,000 meters above sea level, and there the temperature is relatively low, with an annual average temperature of 10-15 °C. Its living environment is also in the high-altitude forest area, which is often cloudy and foggy, and is not be exposed to direct sunlight. However, in the human breeding environment, it is necessary to pay attention to the ambient temperature and light intensity for pandas. Especially in areas with high temperature

and strong sunlight all year round, the strategies of avoiding light, cooling and ventilation must be considered. If pandas suffer from severe heatstroke, emergency cooling measures are required, otherwise, they will die due to heatstroke.

Q: Do giant pandas live in groups?

A: No, they are solitary animals and have their own territory in the wild. Other giant pandas are not allowed to enter, otherwise there will be fights.

Q: Are giant pandas always wandering in the wild?

A: Giant pandas have no fixed place to live, and they do not have a fixed location for eating bamboo. They also poop everywhere, and they are always wandering in their own territory. Giant pandas move and rest randomly according to the bamboo feeding environment and water source distribution. Such strategy greatly saves time and energy consumption when going out for food and returning to the nest.

Q: Do giant pandas have a fixed habitat in the wild?

A: Giant pandas do not have a fixed habitat in the wild. They eat while walking. They like to be alone and wander around. Usually, they do not build nests. Instead, they often live in tree holes, caves or grass piles under trees. They come out in the morning or evening to forage and drink, and for the rest of the time they just sleep. They are relatively docile animals and generally do not actively attack human. They have shorter visual distances and slower movements, but can climb tall trees quickly and nimbly and swim in turbulent rivers.

Q: How is the courtship behavior of giant pandas?

A: The estrus season of giant pandas is from March to May. In the wild, female pandas attract male pandas living around their territory one after another through their calls and the special smell of markers during the estrus season. Usually, male pandas have to fight for several days to obtain the right to mate with female pandas. The best time for a female pandas to conceive is usually only 1-2 days. After mating, the female pandas will drive the male pandas out of their territory and continue solitary life until the baby is born.

Q: When do giant pandas usually get married?

A: The estrus of pandas is single-sided estrus. Its time is mostly in spring, and there are slight differences among different places. When the primroses in the habitat are in bloom, it is their estrus period. It usually starts in late March and extends to mid-May, mostly in mid-April, and in some cases from January to September. In addition, the estrus period of pandas is usually affected by the latitude, altitude and climate of their habitat.

Q: Are giant pandas monogamous in their natural habitat?

A: Giant pandas are not monogamous, but polygamous or polyandrous.

Q: In what season do giant pandas give birth?

A: Under normal circumstances, giant pandas come into estrus and mate in spring, and give birth in autumn, mostly in August to September. Very few pandas give birth until the next spring. Currently, there is only one case in Wolong, Sichuan, China.

Q: How do giant pandas nurture their offspring?

A: The gestation period of a giant panda is usually 3 to 5 months, with a maximum of 10 months. Wild giant pandas generally choose relatively hidden rock caves or tree caves as birth rooms. Newborn cubs are very fragile and need to stay in their mother's arms all the time, and breast milk is their only source of nutrition. The cubs only learn to walk when they are 4 months old, and can eat bamboo at about 1 year old. Therefore, generally, giant panda cubs will live with their mothers until they are about 1.5 to 2 years old, and then they will live independently without their mothers.

Q: What are the physical changes of giant pandas when they grow?

A: Giant panda cubs are flesh-red like little mice when they are born, and their body weight is generally 80-200 grams. After 7 days, black color gradually appeared around the forelimbs, ears and eyes. In about 20 days, the black and white body hair grows evenly, and they gradually gain the ability to regulate body temperature. After about 40 days, the giant panda cub gradually opens its eyes. And their eyes will be fully opened at 2 months, but at this time they are still unable to stand and walk. After more than 3 months, the cub looks exactly the same as the giant panda, and it begins to learn to walk.

Q: How many babies do giant pandas usually give birth to?

A: In the wild environment, giant pandas usually give birth to single young, but in captivity, 50% of giant pandas can give birth to two cubs each time, and rarely give birth to three cubs .

Q: How many babies can a female giant panda have in her lifetime?

A: In the wild, a female giant panda can produce 4 to 8 cubs in her lifetime.

Q: When and where were the wild giant panda twins first discovered?

A: In 1990, the wild giant panda twin cubs, which had been raised for about 45 days, were first discovered in Wolong National Nature Reserve, Sichuan Province.

Q: How long does it take for female giant pandas to get pregnant?

A: The female pandas has displayed delayed implantation, with the shortest being 73 days, the longest being 324 days, and the average being about 120 days.

Q: Why is the gestation time of giant pandas so different?

A: Because the fertilized eggs of giant pandas will go through delayed implantation. After female pandas are mated in estrus in spring, the fertilized eggs reach the uterus from the fallopian tube and do not implant immediately, but after a period of free roaming in the uterus before implanting, the implantation time is delayed by 1.5 to 10 months.

Q: Where do wild giant pandas give birth?

A: Wild giant pandas start looking for suitable caves to make nest before giving birth. In primeval forests, they choose very old

and hollow tree cavities or burrows at the roots of large trees. In the harvested secondary forest where there are few old trees, they choose natural caves or burrows in the forest.

Q: What does newborn baby panda look like?

A: The newly born panda cub looks like a naked mouse, with a relatively big head and a long tail. Its body length is about 15~17 cm, the tail length is 4.5~5.2 cm, the hind feet are 2.2~2.5 cm long, and the weight is about 100 grams. Its eyes are closed, the whole body is pink, with sparse white hair, and the ears are like two fleshy peanuts. It can not stand or crawl, but can raise its head and cry loudly.

Q: What does newborn baby panda eat?

A: Newly-born baby panda only eats its mother's milk. During the first 14 days after the cub was born, it stayed close to her mother, often making shrill cries in its mother's arms. In the next spring, the cub has already begun to walk with its mothers and learn to eat the young bamboo leaves. When its mother is not around, it usually spends some time in the trees, and the tall conifers become a safe island for it to avoid predators. At 8 to 9 months old, the cub starts to be weaned and begins to learn from its mother the skills of feeding in different seasons and different terrains.

Q: Does newborn panda cub grow fast?

A: Newly-born giant panda cub is very small, only 1/1000 of its mother's weight, but it can grow rapidly. A 3-month-old cub can weigh 5-6 kg, and a 6-month-old cub can weigh about 12 kg. When the cub reaches the age from 1 to 4 years old, it can weigh 38, 72, 87 and 97 kg respectively.

Q: What is the average daily weight gain of a juvenile panda?

A: After the cub is born, due to dehydration, poor adaptability to the new environment, and breastfeeding, its weight is always lower than the birth weight in the first 3 days. Its weight can only be restored to the birth weight on the 4th day. From day 5, its weight increases almost linearly with age. Its average weight at 1 month will reach about 1.2 kg, which is about 10 times the weight at birth. It can be seen that the growth and development speed of the cub is very fast. After that, the average daily weight gain will reach about 36 grams in the next month, and 74 to 80 grams in the second to sixth months.

Q: What is the growth of the body length of juvenile panda?

A: The increase in body length is an important indicator to reflect the growth and development of a juvenile panda. Under normal circumstances, its body length is linearly increasing with its age. In general, its length on the first day of birth is 15 cm. It will reach 30.1 cm on the 30th day, 42.1 cm over the 2nd month, 53.2 cm over the 3rd month, 63.0 cm over 4th month, 70.5 cm over the 5th month, and 80.3 cm over the 6th month.

Q: How many parenting postures does the giant panda mother have?

A: For the first three days after the birth of the newborn cub, the giant panda mother nurses the cub in a sitting position most of the time. As it little child grows up, it will increase the supine and side parenting postures, and its strength of the hug also decreases, gradually giving the little child a certain space for movement. In the first few days of her cub's birth, the mother generally cuddles the baby as much as possible, and rarely eats and defecates. Even if it leaves occasionally, the time is short.

Q: What is the purpose of a panda mother licking her cub's whole body with her tongue?

A: There are five main functions: First, it is to clean the whole body of the cub; second, because its saliva contains ferritin, by licking the baby's whole body, it has the effect of anti-disease disinfection and prevention of infection; third, it is to promote the microcirculation of the baby; fourth, it is to keep the skin moist and prevent the excessive evaporation of water; fifth, it is to stimulate the defecation of the baby.

Q: What is the importance of early social activities of giant panda cubs?

A: After decades of observation and research, it has been found that the early social activities of panda cubs will have a huge impact on their future reproductive and maternal behaviors when they turn into adult pandas. Due to the need to increase the population of giant pandas in captivity, the cubs will be separated from the mother by human when the cub is half a year old, so that the mother can be in heat and breed in the next year. However, this practice will make the half-year-old cubs lose the opportunity to live and learn with the mother at an early age, and some cubs do not even have the opportunity to communicate with other cubs. Giant pandas that lack early social activities are likely to have many abnormal behaviors in adulthood, such as strong aggression, incorrect mating posture, and incorrect feeding methods for their newborn babies. Relatively speaking, the lack of early social activities of these young cubs has greater and more significant adverse effects on male pandas.

Q: How does a panda cub communicate with its mother?

A: Before the panda cub has any hearing and vision, it mainly relies on their sense of touch, smell and calls to communicate with their mother, and feedback information such as heat and cold, hunger, tightness of the hug, and whether it is looking for nipples. Its mother will judge the different needs of her cub according to the cries. Sometimes she will move the cub to get close to the nipples, sometimes she will lick it, and sometimes she will walk around with her cub in her mouth. In a word, the mother will adjust her behavior at any time according to the needs of her cub, so the first three days after the birth to the cub will be very hard for her.

Q: What are the behaviors of newborn panda cubs?

A: The newborn panda cubs are not fully developed in the mother's body, and their sensory systems have not been fully grown. The cubs and their mothers mainly rely on sound to communicate. Their common behaviors are:

(1) **Screaming** Their screaming is divided into two situations, one is a low pitch sound, and the other is a high pitch sound. This sound is made when the cubs feel cold outside of the mother, usually to ask the mother to adjust her position.

(2) **croaking** This kind of call is similar to frog croaking, but the pitch is very low. It is often made immediately after the scream, indicating a pleasant feeling. It is the signal sent by the cub when the mother is adjusting its position and the cub feels comfortable.

(3) **Continuous screaming** This is a signal sent by the cub when it feels extreme discomfort or when it asks the mother for breastfeeding. Sometimes such scream will continue for a long time until the cub feels satisfied.

(4) **Trembling** When the cub is exposed outside the mother, the cub will tremble because of the coldness. Then, the mother will adjust the position of the cub and completely cover it to keep it warm.

(5) Breastfeeding When the cub is hungry, it will look for the teat and make continuous screams. At this time, its mother will help it to find the teat. Before it reaches 1 to 5 days old, the cub feeds at different intervals, 6 to 12 times a day and night. The duration of each feeding varies from half a minute to more than 10 minutes, and some as long as 30 minutes. After it reaches 15 days old, the frequency of breastfeeding gradually decreases to 3 to 4 times a day.

(6) Smell As the cub grows, it begins to explore the surrounding environment by smell, and the communication between mother and the cub also begins in this way.

(7) Crawling Before the cub reaches 2 months old, it has almost no ability to move, except for feeding or sleeping, it can only wriggle when looking for nipples. After 2 months old, it can crawl, but its limbs are not coordinated and the crawling distance is short.

(8) Unsteady walking At around 4 months old, the cub starts to walk, but the strength of its limbs is not balanced, the walking is unstable, and it often rolls on the ground.

(9) Walking and climbing When the strength of the cub's limbs is balanced, it will start to acquire the ability for normal activities. At this time, it is able to not only walk around but also climb, but the range of these activities is not long. Sometimes when its range of activities gets too long, its mother will bring it back.

Q: It is said that panda mothers sometimes abandon their cub, is that true?

A: When female pandas give birth to one cub, they will take good care of it; when they give birth to two or more cubs, almost all female pandas will choose the healthiest and strongest one to feed, and discard the rest ones. This is the abandonment behavior of giant pandas. In captivity, some female pandas do not have the ability to care for all their cubs, and the abandoned cubs are usually fed by human.

Q: Does inbreeding affect the survival of wild giant panda populations?

A: All species have developed their own unique mechanisms and strategies to avoid inbreeding. So far, there is no direct evidence showing that inbreeding is a problem for wild pandas. But the fact is that in some areas, the giant panda habitat is severely fragmented, and giant pandas are fragmented into multiple small populations. Future conservation work will focus on connecting these isolated populations, make them communicate with each other so as to improve genetic diversity.

Q: How long does it take for giant pandas to start teething after birth? When does a cub develop its baby teeth?

A: A giant panda cub begins to grow teeth 3 months after birth, and by 6 months, its baby teeth are basically fully grown.

Q: When does a panda start changing its teeth? How long does it take for its teeth replacement process to end?

A: When the panda cub grows to about 8 months old, the deciduous teeth gradually fall off and its teeth replacement process begins; all the deciduous teeth are replaced by the permanent teeth at the age of 15 to 17 months.

Q: When are giant panda cubs weaned?

A: In the wild, from birth to 7 months old, the nutritional acquisition of the cubs is completely dependent on the mother's breastfeeding, and the weaning begins at about 8 to 9 months old.

Q: Will male panda help to raise its cub?

A: The male panda will not help the female panda to raise their cub. It leaves the female panda after mating. The female panda undertakes all the pregnancy, calving, raising and training of the cub.

Q: How old do wild giant panda cubs leave their mothers to live independently?

A: Wild giant panda cubs usually leave their mothers to live independently at the age of 1.5. The daughter will be married to other places, but the son still lives in its mother's nest. It will leave its mother's nest at the age of 2.5 to establish its own territories.

Q: What is the importance of learning for the panda cubs?

A: Along with their own growth, giant panda cubs will learn many skills from their mothers, such as avoiding predators, climbing trees, choosing appropriate bamboo, and finding water sources. In the wild, when a panda cub reaches 1.5 to 2.5 years old, it will have to face the world independently. Typically, after a cub has grown up, it has to forage, play, and rest alone. However, in the early days of its independent life, the young panda usually does not get too far from its mother's nest, and will not establish its own territory of activity until its gradually adaption to the outside world.

Q: Is it easy to see giant pandas in the wild?

A: Giant pandas are called "hermits in the forest". Because giant pandas live in lush and dense bamboo forests and are often solitary, it is not easy to see individual giant pandas in the wild. But if you enter the forest area where giant pandas live, it is easier to see their life traces, such as their feces and chewed bamboo joints.

Q: What other sympatric or companion species are there in the giant panda habitat?

A: The companion species of giant pandas include golden monkey, takin, red panda, black bear, wild boar, green-tailed rainbow pheasant, red-bellied pheasant, forest musk deer, bamboo rat, macaque, hairy deer, wildebeest, blood pheasant, Golden pheasants, etc. Theses precious animals basically live in the same environment as giant pandas. They share a common evolution experience and have been attacked and severely tested by glaciers, but they have survived to this day with the help of the favorable terrain of high mountains and deep valleys in the southwest of China. Through long-term coexistence and coordinated co-living in the same area, they each occupy their own space and form a relatively stable community.

Q: What are the main natural enemies of giant pandas?

A: Giant pandas can often live in harmony with other wild animals, but frail, sickly or young giant pandas are vulnerable to predators such as leopards, jackals, wolves, and lynx.

Q: Should giant pandas be called "daxiongmao" (Chinese: big bear cat) or "damaoxiong" (Chinese: big cat bear)?

A: The latter one actually is a more suitable name for giant pandas in Chinese, because they are close relatives of bears, but distant relatives of cats. It was an old "mistake" caused by a "Westernization" change in the sequence of reading and writing of Chinese written language in mainland China in the 20th century. Modern Chinese have also recognized this name, making the pandas "like a bear but not a bear". Even the most critical taxonomists recognize this mistake" because they are so unique, and special. Yet, the name of "maoxiong" (Chinese: cat bear) are still preserved in books and periodicals in Taiwan Province, China.

Q: Are panda paws the same as bear paws?

A: The forefoot of the giant panda is different from that of other bears. The forefoot of the black bear is relatively long and narrow, and the middle finger is also long and protruding. Its claws are long and sharp. The forefoot of the giant panda is slightly round, the five fingers are basically aligned, and there are hairs between the fingers.

Q: What is the difference between a giant panda and a black bear?

A: A black bear looks slender, with all black coat, only with a patch of white hair on its chest. It has slender front paws, with sharp claws. It also has a long head, with and a long snout. The black bear belongs to omnivore with 37 pairs of chromosomes. The giant panda is round in shape, with black limbs, ears and eye rims, and white other parts. Its forefoot is oval, the five fingers are neat, and there is a sixth "pseudo thumb". A giant panda has a round head with a short snout. 99% of a giant panda's food is bamboo. A panda's premolars are large and flat. It also has 21 pairs of chromosomes.

Q: How is the skull of the giant panda different from other bears?

A: Giant pandas are highly specialized carnivores. Although their appearance is similar to that of bears, many of their anatomical features are different from those of bears. Their skull and related muscles are obviously different from other bears. The masticatory muscles are well developed, and the grinding surfaces of the premolars become larger. These changes are related to the high dependence on eating bamboo.

Q: How are the teeth of giant pandas different from other bears?

A: From an anatomical point of view, the tooth shape of giant pandas is different from other omnivorous bears. The premolars of giant pandas are flat and wide, which can help them to easily bite and grind bamboo.

Q: Are giant panda and red panda different species?

A: The giant panda and the red panda are different species. In June 1825, a French zoologist called Georges-Frédéric Cuvier discovered the red panda in the Himalayas. He named it

Ailurus fulgens and called it "Panda" in English. Because giant panda was discovered later, the English name for red panda was then called Lesser Panda or Red Panda to distinguish them. In Chinese, its name was translated as "little panda", and Giant Panda was "big panda". The red panda is very beautiful in appearance. It is about the size of a fox, with a brown-red coat. It has a round face, white ears, cheeks, eyes and mouth, and nine white rings on its long and fluffy tail hair. That's why local people call it "Nine-section Wolf".

Red pandas are distributed in Nepal and Qinghai-Tibet Plateau of China, and some of them live within giant panda habitats. Red pandas like to move on trees. They only go down to the ground when foraging. They also like to eat bamboo leaves, and also eat wild fruits, bird eggs, small animals, etc.

Q: Do pandas have many anthropomorphic movements?

A: The cuteness of giant pandas is that they have many behaviors and habits in common with humans. For example, giant pandas can stretch their waists and often sit and play. It will also stand up if its line of sight is blocked or when it needs to get objects from high above. When sleeping, if the light is too bright, they will cover their eyes with their forefoot and take a nap. After waking up, they will also yawn one or more times, showing a very pleasant expression. The sound of them eating bamboo shoots, wotou, fruits and other foods, which sounds like "Hey ah... ah ah... ah ah...", is the same as when people eat very delicious food. When the pandas are eating bamboo poles, the feeling of "clear, crisp and refreshing" is quite obvious to observe, which will even make people to drool.

Giant Pandas' Home in the Wild

Q: Where are the type specimens of giant pandas produced?

A: The type of origin refers to the geographical location of the original specimen used to name the species. The type specimens of giant pandas are produced in Dengchigou, Baoxing County, Sichuan Province, China.

Q: Where was the giant panda's former home?

A: Historically, giant pandas once lived in subtropical evergreen forests covering where is now most of China, Laos, northern Myanmar, northern Thailand, and northern Vietnam. Due to the massive loss and fragmentation of habitats caused by human reclamation, logging and roads building, the giant panda population has retreated to China's mountainous areas like the southern foot of Qinling Mountains in Shaanxi Province, Minshan Mountains, Qionglai Mountains, Daxiangling Mountains Xiaoxiangling Mountains and Liangshan Mountains in Gansu and Sichuan provinces.

Q: How big is the giant panda's habitat?

A: The results of the fourth national giant panda survey released in 2015 showed that wild giant pandas are distributed in 17 cities (prefectures) and 49 counties (cities, districts) in Sichuan, Shaanxi and Gansu provinces, and the total habitat area has expanded to 2.58 million hectares, and the wild panda population has increased to 1,864. Due to factors such as natural isolation or human disturbance, the wild population of giant pandas has been divided into 33 localized populations, of which 18 populations have fewer than 10 pandas and are at high risk of extinction.

Q: What is the wild home of giant pandas like?

A: Giant pandas are animals living in temperate zones. They generally live in a mixed environment of broad-leaved forests and sub-mountainous dark coniferous forests at an altitude of 2000-3700 meters. There are lush bamboo forests under the forest layers, serving as their main food. In addition, giant pandas like to move on relatively gentle hillsides, which can reduce energy consumption; there are often streams near their living areas where they can drink water. Since the giant panda is a warm-blooded animal, the altitude at which it lives varies with the seasons.

Q: What is the climate of the giant panda distribution area?

A: The climate in the distribution area of giant pandas is a typical warm and humid mountainous climate, with no extreme heat in summer and abundant rainfall.

Q: How big is the giant panda's territory?

A: In the wild, giant pandas of different genders and ages have different areas and sizes of territories due to human factors, forest coverage and the amount of edible bamboos. The area of male giant pandas is generally slightly larger than that of females, about 6 to 7 square kilometers; the area of sub-adults is generally 4 to 6 square kilometers.

Q: How long does it take for a large-scale bamboo flowering to occur in the giant panda habitat?

A: Bamboo flowering refers to the sexual reproduction phenomenon of bamboo when it blooms in a large scale every 30 to 120 years. One or several kinds of bamboos in the giant panda habitat wither and die after flowering in a large scale. It will take several years of seed recovery before they gradually germinate and grow again. After the bamboos bloom and die, it usually takes a period of nearly 10 years for the bamboo population to recover. During this recovery period, giant pandas must find other bamboos that have not bloomed for food.

Q: Will large-scale bamboo blooms in giant panda habitats threaten the survival of panda populations?

A: Since giant pandas mainly rely on bamboo for their living, this natural phenomenon within the bamboo growth cycle brings certain survival problems to giant pandas. When a type of bamboo blooms in a large area, it will increase the difficulty for pandas to obtain food. For example, in the mid-1980s, large-scale bamboo blooms caused the death of many giant pandas. However, inside their habitat, there are usually at least two types of bamboo. When the food supply of one type of bamboo is reduced due to flowering, giant pandas can migrate to feed on the other type of bamboo, or expand their territory and feed on non-flowering bamboos. However, the giant panda's habitat has been fragmented due to the interference of human activities, and the migration of giant pandas between different habitats has been blocked. Therefore, large-scale bamboo flowering is also a potential threat to the survival of the current panda population.

History of Giant Pandas

Q: When did the western world begin to know about giant pandas?

A: March 11, 1869.

Q: Who was the first Westerner to discover giant pandas? What is his occupation?

A: A Frenchman called Armand David. He was both a missionary and a taxidermist working for a museum.

Q: Who was the first to take a live giant panda out of China? Is she a zoologist?

A: Her name is Ruth Harkness. She is neither a zoologist nor a zoo employee, but an American costume designer.

Q: Did Ruth name the captured panda cub "Su Lin"?

A: Ruth mistook the panda cub as a female, so she named it after the wife of the Chinese accompanying her.

Q: How long did the first live giant panda live after it arrived in the United States?

A: It lived in the United States for just over a year and died in April 1938.

Q: How were giant pandas discovered?

A: When it comes to the "discovery" of the giant panda, the French missionary David must be mentioned. He believes that he discovered this "strange" animal in 1869, which was difficult for the Western world to recognize at that time. Many people think that it is impossible to have such a uniquely "conceptualized" creature in the world. Its huge body is like a bear, but with distinct black and white color; a round white head with two black eye sockets and black ears; black front and rear legs are clearly separated by a white back and belly; almost tailless. To be clear, for Father David's "discovery", a more accurate statement should be that he introduced the magical species of giant panda to the West. The panda was given a scientific name by scientists who were convinced of its existence in 1870 according to the provisions of modern taxonomy. In fact, Chinese people have known about the existence of giant pandas as early as 3,000 years ago in the Western Zhou Dynasty, but they were not called giant pandas at that time, but "*Pixiu*". Their specific distribution was recorded in China's geographical work called *Classic of Mountains and Seas*: "Like a bear, black and white beast... lives in the south of Yandao County, Qionglai Mountain". Since there was no systematic classification in ancient China, and there were many common names in various places, giant pandas had a long list of names in Chinese history.

About Giant Panda's Conservation

Q: What work has China done in *in situ* conservation of giant pandas?

A: The Chinese government has issued a number of laws and regulations such as the *Wildlife Protection Law of the People's Republic of China* and *the Regulations on the Management of Nature Reserves of the People's Republic of China*, and implemented key forestry priorities such as natural forest resource protection, returning farmland to forests and grasslands, wildlife protection and the construction of nature reserves. The project will continuously improve the nature reserve system. Today, the conditions of the giant panda have been significantly improved, the wild population has gradually recovered, and the number of panda's nature reserves has reached 67, with a total habitat area of 2.58 million hectares. Four national giant panda surveys have been carried out.

Q: What are the three main approaches to the conservation of endangered giant panda species?

A: The first approach is *in situ* conservation, which is the ideal method to protect its habitat and allow it to reproduce naturally. The second is *ex situ* conservation. Zoologist from zoos and giant panda scientific research institutions will breed and feed giant pandas to expand their captive-bred population. The third is the hybrid approach by combining the first and the second methods. While protecting panda's natural habitat, the captive-bred panda will go through "wild survival training" and then be released to the habitat, so as to enrich and expand the wild panda population.

Q: How many giant panda nature reserves are there in China?

A: So far, 67 giant panda nature reserves have been established.

Q: Which giant panda nature reserves were established first?

A: China issued the "*Instructions on Active Protection and Rational Utilization of Wild Animal Resources*" in the 1960s, and for the protection of wild giant pandas, the first four giant panda nature reserves were designated in 1963. It is the Wolong Nature Reserve, Baishuihe Nature Reserve, Wanglang Nature Reserve and Labahe Nature Reserve, covering an area of about 900 square kilometers. After 1970, according to the distribution of giant pandas, 8 national nature reserves were further established: Sichuan Fengtongzhai National Nature Reserve, Shaanxi Foping National Nature Reserve, Gansu Baishuijiang National Nature Reserve, Sichuan Mabian Dafengding National Nature Reserve, Sichuan Tangjiahe National Nature Reserve, Sichuan Xiaozhaizigou National Nature Reserve, Sichuan Meigu Dafengding National Nature Reserve, Sichuan Jiuzhaigou National Nature Reserve. In the 1980s, 23 new nature reserves were further established to protect giant pandas.

Q: When did China's earliest scientific research activities on giant pandas begin?

A: In 1974, led by Professor Hu Jinchu from the Department of Biology, Nanchong Normal University, Sichuan (now the School

of Life Sciences, Xihua Normal University, Sichuan), a field survey and research team of about 30 people, which was named as Sichuan Rare Animal Resources Survey Team, was established. They started the first wild giant panda survey in China (and also the world's first comprehensive census of the number of wild giant pandas).

Q: When and where was the first giant panda field observation station built?

A: The first field observation station for giant pandas, the "Wuyipeng" field observation station for giant pandas, was established in Wolong National Nature Reserve in Sichuan in 1978. It was the first giant panda field observation station.

Q: How did the "Wuyipeng" get its name?

A: The field observation station was built with canvas back then. There is a staircase connecting the tent station and the water intake pool, with a total of 51 steps. Scientists call this observation station built with canvas tents "Wuyipeng" (Chinese: a tent with 51 steps).

Q: What is the altitude of the first giant panda field observation station?

A: 2,520 meters.

Q: What other observation stations are there for giant panda ecology research?

A: There are also Foping Sanguan Temple Observation Station and Yangxian Giant Panda Observation Station in Shaanxi Province, Tangjiahe Baixiongping Observation Station and

Mabian Dafengding Observation Station, Mianning Yele Observation Station in Sichuan Province, etc.

Q: When was China's first institution specializing in giant panda conservation research established?

A: With the support of the World Wide Fund for Nature (World Wildlife Fund International), China established the "China Conservation and Research Center for Giant Panda" in 1981. Since then, international cooperative research on giant pandas has been initiated. The center has also played an important leadership role in the *ex situ* and *in situ* protection work of giant pandas.

Q: When was China's project to protect giant pandas and their habitats launched?

A: In 1992.

Q: When were the four special field surveys on giant pandas organized by the Chinese government?

A: They are: from 1974 to 1977, the number of giant pandas found in the first survey was about 2,459; from 1985 to 1988, the number found in the second survey dropped to more than 1,114; from 1999 to 2003, the number found in the third survey was 1,596; from 2011 to 2014, the number found in the fourth survey reached about 1,864.

Q: What are the methods of surveying giant pandas in the wild?

A: There are several methods for surveying giant pandas in the wild, such as:

Bite knot method: According to the length, thickness and chewing degree of bamboo knots in the feces of giant pandas, the approximate age, individuals, population numbers, and activities of giant pandas can be identified.

DNA identification method: collect giant panda feces or hair and other tissue samples in the wild, and extract panda's DNA for analysis. Nowadays, there are several mature analytical technologies, which can more accurately determine the identity, age, gender, blood relationship, etc. The disadvantage is that the technology applied is complicated and the cost is high.

Footprint recognition method: Dr. Li Binbin of Duke Kunshan University announced at the end of 2017 that he has established a footprint model recognition method that can be used to identify individual giant pandas and their gender. This method still requires field practice to verify its effectiveness.

Q: When was China's Giant Panda National Park established?

A: The Giant Panda National Park was officially established in October 2021. It spans Sichuan, Shaanxi and Gansu provinces with a total area of 22,000 square kilometers. It is the core distribution area of wild giant pandas.

Q: What is China's Giant Panda National Park like?

A: The Giant Panda National Park is located in the boundary line between the first and the second geographical step in China, where the transition zone from the eastern edge of the Qinghai-Tibet Plateau to the Sichuan Basin lies, and here the terrain is complex. The mountains are high and the valleys are deep, and the water system is well developed. The lowest altitude here is 529 meters, the highest altitude is 6,250 meters, and it is very common to see deep valleys with a relative height difference of more than 1,000 meters. The river system belongs to the three water systems of Jialing River, Minjiang River and Tuojiang River in the Yangtze River Basin. It is recognized as one of the most complex geomorphic areas in the world, and it is also an area with frequent geological disasters.

Q: What's in the China's Giant Panda National Park?

A: The Giant Panda National Park has integrated 73 nature reserves that originally belonged to different departments or administrative regions. To enhance the connectivity of 13 local populations in the National Park and increase gene exchange, we strengthened habitat renovation and restoration and corridor construction work. There are 1,340 wild giant pandas in the Giant Panda National Park, accounting for 71.89% of the entire wild giant panda population. The geographical topography and ecology of Giant Panda National Park are complex and diverse. There are many rare wild animals and plants such as golden leopard, snow leopard, Sichuan golden monkey, forest musk deer, takin, yew, dove tree, etc., and the biodiversity is very rich.

Q: Why do we rescue wild giant pandas?

A: Carrying out rescue work for sick, hungry and injured giant pandas in the wild can greatly reduce the excess mortality of wild giant pandas. Through our rescue and recovery work, the rescued giant pandas in good physical condition are released back into the wild, which can effectively ensure the population of wild giant pandas. This important function has long been shouldered by China's giant panda nature reserves, forestry authorities at all levels, and giant panda breeding and breeding institutions. So far, nearly 300 giant pandas have been rescued, and giant pandas such as "Si Guniang", "Sheng Lin No. 1" and "Lu Xin" have been successfully released into the wild.

Q: How are wild giant pandas discovered after being injured or sick?

A: After being injured or sick, wild giant pandas often move to lower altitudes and are discovered by conservation personnel or villagers patrolling the mountains.

Q: How to deal with wild sick and hungry giant pandas after rescuing and curing them?

A: After rescue and cure, individual pandas who are capable of returning to the wild will be released into the wild; those who are unable to return to the wild will be kept in captivity.

Q: What have humans mainly done for the rejuvenation of the giant panda population?

A: ① Legislation to enforce mandatory protection of giant panda species. ② Established protected areas to protect their original ecological environment and habitats. ③ Implemented the natural forest resource protection and the project of returning farmland to forest to protect and expand panda's habitat. ④ Established a green "corridor belt" of pandas, so that small populations in isolated communities can move smoothly with each other, forming larger breeding populations, creating conditions for gene exchange, making populations rejuvenated, and preventing the loss of their genetic diversity. ⑤ *In situ* conservation, carrying out multidisciplinary research on giant pandas. ⑥ Trial research on the release of captive giant pandas into the wild.

Q: What are the differences between *in situ* conservation and *ex situ* conservation of giant pandas?

A: From the perspective of the overall protection of giant pandas, *in situ* protection mainly focuses on habitat protection, while *ex situ* protection focuses on the management of panda's breeding and population; *in situ* protection has a large scope, lots of contents, and very complex issues to manage, while the issues of *ex situ* protection is simple, and the research content is relatively concentrated; the environment of *in situ* protection is poor, basically in the deep mountains, wages and benefits are not attractive, and the brain drain there is serious; while for *ex situ* protection, its location is usually close to large cities, and the source of funds is diversified, which can retain highly educated and high-level talents. Therefore, in order to maintain the high level momentum for the giant panda protection work, we must attach great importance to the introduction and training of local protection talents, and at the same time, we must give corresponding treatment and honor to field workers, care and love, and fully realize that talents are the key to promoting giant panda protection.

Q: What work has the Chinese government done in *ex situ* conservation of giant pandas?

A: Through the long-term and unremitting efforts of giant panda researchers in China and abroad, great progress has been made in the research of giant panda biology, physiology, genetics, ecology, conservation science, medicine and other aspects, and the international challenge of giant panda breeding has been overcome. By the end of 2021, the captive population of giant pandas has reached 673; 9 captive-bred giant pandas have been released into nature and successfully integrated into the wild population.

Q: What is the effect of *ex situ* conservation of giant pandas?

A: *Ex situ* conservation of giant pandas is an important supplement work to *in situ* conservation. The International Union for Conservation of Nature (IUCN) advocates that when the total number of an animal population in the entire natural environment drops to about 1,000 (heads), it is necessary to transfer some of these rare animals to suitable, safe and secure artificial environments. Through rearing and breeding in captivity, the population of this species can achieve self-reproduction and maintenance. After reaching a certain number, it will be released into the wild to rebuild and rejuvenate its wild population in a planned and scientific way.

In addition, in the process of *ex situ* conservation of giant pandas, it is convenient to conduct multidisciplinary scientific research on giant pandas to gather and analyze data on their behavior, habits, physiology, and diseases. A variety of public education activities can also be carried out to raise people's awareness of protecting giant pandas and their ecological environment.

Q: Why is breeding giant pandas one of the world's problems in the 20th century?

A: First, in the 20th century, the number of male giant pandas that could participate in breeding was rare. At that time, there were about 120 male pandas in China and abroad, and only one was able to participate in natural mating. Second, it is difficult for male pandas to come into estrus. Female pandas entering the breeding season in captivity can generally come into heat on a seasonal basis, but males are difficult to come into heat, and artificial insemination was also used as a supplementary method at that time. But female giant pandas can conceive only about three or four days in a year, and there are no obvious physiological characteristics. Therefore, in artificial insemination, it is difficult to accurately grasp their ovulation period. For example, in 1989, 18 female giant pandas were either artificially inseminated or mated naturally, some were even given both methods, but only 6 were conceived. Third, it is difficult for the cubs to survive. A newborn cub is only the size of an ordinary mouse and weighs about 100 grams, one-thousandth of the weight of an adult panda. In mammals, except marsupials, there is no animal with such a disparate size ratio between mother and child. The cubs have closed eyes, can't see anything, and have only sparse lanugo on their skin. Their level of development is about the same as that of a six-month-old human embryo, and their kidneys, brain and some immune organs and lymphoid tissues are not fully developed. This kind of organism may have adapted to the environment of high mountains, ice and snow, where there were rarely any bacteria, but it cannot adapt to the *ex situ* environment. Giant pandas have low mating rate, low conception rate and low survival rate, and these are their particular challenges among captive animals. This "three lows" was the problem of breeding giant pandas at that time.

Q: How has China overcome "the three major challenges" in the breeding of captive giant pandas?

A: After more than 30 years of hard work, China has made great achievements in the breeding of captive giant pandas, overcoming the "three major difficulties" (i.e., difficulty in estrus, difficulty in breeding and conception, and difficulty in nurturing and surviving) and achieving stable progress. The first difficulty is overcome through careful and responsible feeding, improving nutritional supply, and providing animal welfare, exogenous hormone induction and behavior induction; The second difficulty is overcome through breeding male pandas, ovulation and pregnancy monitoring, panda semen collection and preservation, and panda insemination; The third difficulty is overcome through "mother and cub" training, improving technology in cub breeding and artificial formula panda milk.

Q: Why artificial insemination?

A: Artificial insemination is an effective method to improve the conception rate of giant pandas under captive conditions. Researchers keep an eye on female panda's behavior and hormonal changes to grasp the optimal mating time, collect the semen of male pandas, and artificially inseminate those who cannot mate or are not ideal for natural mating. Artificial insemination can increase the pregnancy probability of female giant pandas and provide more options for achieving genetic diversity.

Q: Why do human help to babysit panda cubs?

A: Giant pandas generally give birth to 1-2 cubs per litter, and there are also rare cases of 3 cubs. But usually, a female panda is only able to nurse one cub, and some doesn't even nurse at all. The researchers carried out human babysitting work by imitating panda mother's parenting environment, developing artificial milk, and exchanging cub rearing, which ensured the survival, normal growth and development of most giant panda cubs.

Q: Why do we promote prenatal and postnatal care for pandas?

A: Because if we take measures like strict blood and genetic management which can help us avoid inbreeding, removing older and less physically fit giant pandas from breeding, and extending the time cubs spend with their mothers to learn more life skills, it is possible to effectively improve the individual quality of captive giant pandas, maintain genetic diversity, and improve their population vitality.

Q: When did China start to raise giant pandas in captivity?

A: As an very ancient creature still living in nature, giant pandas have a very long history of captive breeding in China. *According to the Records of the Grand Historian*, Xuanyuan Huangdi gathered tapirs, tigers and other beasts to fight against Emperor Yan in a place called Quan. This is the evidence showing that captive giant pandas were used for war long ago. According to Sima Xiangru's *Shanglin* Fu, during the Western Han Dynasty, there were many giant pandas in captivity in the Royal Shanglinyuan Animal Hunting Ground near Xianyang City, Shaanxi Province, which is the earliest record of captive bred giant panda in the world. During the Tang Dynasty, two living giant pandas and 70 panda furs were given to the then Japanese Emperor Tenmu. It is the earliest record of living giant pandas outside China.

Q: Where did the first giant panda in captivity come from after the founding of the People's Republic of China?

A: On January 17, 1953, a giant panda cub was successfully rescued in Baimagou, Yutang Town, Dujiangyan City, Sichuan Province. As the first institution to raise giant pandas in captivity after the founding of the People's Republic of China, Chengdu Zoo named this cub "Da Xin".

Q: When was the first captive-bred giant panda cub born?

A: On September 9, 1963, the giant panda "Ming Ming" was born at the Beijing Zoo, becoming the first giant panda bred in captivity.

Q: When was the world's first successfully raised captive giant panda twins?

A: On August 24, 1990, a panda named "Qing Qing" in Chengdu Zoo successfully bred its first panda twins. It was also the world first human bred panda twins, making a major breakthrough in the history of giant panda breeding. Because she was born on the eve of the Beijing Asian Games, it was named "Ya Ya" "Xiang Xiang", and has attracted much attention both in China and abroad. The successful breeding and survival of the giant panda twins is a clear demonstration of China's reserve in research and technology in the domain of panda breeding, dawning a new era of captive breeding of giant pandas.

Q: What is the nutritional composition contained in the food of captive giant pandas?

A: The main food of adult giant pandas is definitely bamboo, which accounts for about 95% of their total food. Giant pandas have no freedom to choose the species of their bamboo if the food is prepared by humans. Therefore, to ensure their nutrition intake, they must be fed with some selective feed. In this way, it is possible to avoid insufficient nutritional intake. The selective feed are mostly composed of plant-based raw materials, such as wheat, soybean, corn, rice, wheat bran and so on. Nutrition wise, the selective feed is roughly composed of 6% to 17% of protein, 3% to 4% of crude fat, and 75% of carbohydrates. Also, it requires some fruits and vegetables as supplement.

Q: Does bamboo need to be rinsed before feeding?

A: Bamboos retrieved from storage will somehow be contaminated, but they can be cleaned just with clean tap water without disinfection.

Q: How to control the total amount of bamboo fed to giant pandas every day?

A: The total daily amount of bamboo depends on the size and food intake of the panda. According to calculations, the amount of bamboo eaten by the giant panda is about 6% to 15% of its body weight. Generally, for each adult giant panda, it is necessary to prepare 2 to 3 times of feeding for each day, sometimes including night.

Q: What are the specific requirements for the playgrounds for captive giant pandas?

A: The area of the playground is generally required to reach 300 to 500 square meters, the absolute height of the partition wall around the playground should not be lower than 2.8 meters, and the surrounding walls should be smooth without gaps and protruding points. If we consider tourists to visit, it is necessary to dig a 1.2-meter-deep trench, with its upper wall covered by 1.2-meter-high smooth glass to isolate the giant pandas. Then, deploy anti-escaping power grids around the trench and separation wall, set the pulse voltage to 700-1,000 volts or 5,000-10,000 volts, and keep the power on for 24 hours to form a double insurance of safety. In the daytime or when there are keepers, the power can be set to low-voltage pulse mode; at night or when the alert level is high, it can be adjusted to high-voltage pulse mode.

The enrichment of play items as well as plantings in the playground is also an important part that cannot be ignored. For the terrain design, it is best to have an undulating terrain. For its planting, 5 to 10 trees with large crowns should be planted, and some shrubs should be included. Use lawn mulch which will live well under local climatic conditions to cover the ground.

It is best not to build high walls around to avoid air flow block. In addition, water supply and drainage should be considered. A 4-5 square meter pool can be built with stone in the central area of the playground. The pool can be used as a drinking and bathing spot for giant pandas, and a waterfall can be designed to flow down from a high place. The drainage in the playground should be unobstructed, the light and dark ditch should be combined, and the setting of the bamboo residue filter should be considered to avoid the blockage of the drainage system.

Q: What does the "environmental enrichment" of captive giant pandas mean?

A: Locking a relatively free animal in a single, limited environment may be a good sanctuary for an endangered animal, but from the perspective of animal habits, it will lose a lot of natural characteristics and freedom. This has put forward a demand for animal breeding management agencies to enrich the environment as much as possible when raising giant pandas in captivity. It should enable giant pandas to avoid strong sunlight and bathe in mild sunlight; they can both walk on the ground and climb trees; they have both open running space and hidden places, which can avoid some human stimulation and interference.

Q: How to identify different individuals of captive giant pandas?

A: The breeders of giant panda breeding institutions are all able to identify the giant pandas they have cared for many years. They mainly distinguish different individuals based on their age, body size, facial features, coat depth and other characteristics. In addition, tattoo numbers are marked on the mouth or skin of most giant pandas, and electronic tags are embedded under the skin of giant pandas, and the archives of giant pandas can be read through a special electronic card reader. According to the latest news, researchers have also successfully developed a panda facial recognition system, which will make the individual identification of giant pandas easier and more convenient in the future.

Q: What is the function of the giant panda radio collar?

A: First, after monitoring at more than three positioning points, you can know exactly where the giant panda is and whether it is moving. Using this method, you can find out the regularity of the giant panda's movement. Much information shows that giant pandas generally live alone, not fixed in one place, but have a small range of activities, feeding, wandering and resting for a short period of time around the clock. Second, in different seasons, through the abnormal changes in the pulse frequency of radio signals, we can learn about important events like the estrus, mating and reproduction of giant pandas. Third, the population structure of giant pandas can be further studied.

Q: What are the difficulties in releasing captive-bred giant pandas into the wild?

A: The main difficulty is to teach giant pandas how to establish their own territory in the wild, how to choose food to meet their nutritional needs, how to recognize and avoid natural enemies, and how to deal with external parasites.

Q: Why are giant pandas released into the wild?

A: The human bred giant pandas will be released into the wild after wild training, so that they can be integrated into the wild panda population, improve the genetic diversity of wild pandas in local areas, increase the number of wild pandas, and improve their survival capability. These are the major purposes of carrying out artificial breeding, breeding and scientific research of giant pandas.

GIANT PANDA! 大熊猫 201 问

Q: What is the first step in the release of captive-bred giant pandas into the wild?

A: Captive giant pandas grow up under the care of human beings, and they are relatively dependent on humans. Under captive environment, we must first ensure the health of giant pandas, and then try to reproduce as much as possible, so as to increase the number of giant pandas as soon as possible in a short period of time. While pursuing quantity, if the improvement of the overall quality of giant pandas is ignored, giant pandas will lose many natural behaviors under human intervention and in restricted environments. Restoring their lost behavior is a very difficult task. Relatively speaking, it is easy to lose something, but very difficult to restore it. Creating conditions to restore the lost behavior of giant pandas has become an important starting point in their reintroduction process.

Q: What goals must be achieved for a successful panda wild reintroduction?

A: After long-term wild training, the captive giant panda can be restored to its wildness survival ability, and released into nature. If a panda can adapt to the wild environment and live a normal and healthy life, it means it has reached the first step in reintroduction into the wild. In natural environment, if giant pandas released into the wild can live in harmony with their surrounding partners and companion animals, and can avoid the threat of individual predators and the invasion of parasites and diseases, it means their survival ability have reached a higher level. Not only that, the ultimate goal of the wild reintroduction is that, through competition in nature, a panda can find one's favorite mate, successfully mate, successfully give birth and nurture, and enter a state of continuous reproduction.

Q: In which year was the first training of captive giant pandas released into the wild? Where did it take place?

A: On July 8, 2003, in Hetaoping, Wolong Nature Reserve.

Q: When was the giant panda "Xiang Xiang" officially released into the wild?

A: April 28, 2006.

Q: What is the name of the cub that has received the first "mother and cub" training under our giant panda wild reintroduction effort and been released into the wild afterwards?

A: This giant panda is called "Tao Tao".

Q: How many giant pandas have been released into the wild? How are they doing in the wild?

A: China started to release giant pandas into the wild in 2006. By the end of 2018, captive bred giant pandas have been released into nature 8 times, and there were 11 pandas released.

The time are as follows:

April 2006, "Xiang Xiang" ;

October 2012, "Tao Tao";

November 2013, "Zhang Xiang";

October 2014, "Xue Xue";

November 2015, "Hua Jiao";

October 2016, "Hua Yan" and "Zhang Meng";

November 2017, "Ba Xi" and "Ying Xue";

November 2018, "Qin Xin" and "Xiao Hetao";

Among them, "Xiang Xiang" and "Xue Xue" failed to survive. Based on the collected monitoring data, experts determined that the other 9 pandas have achieved the staged goal of survival in the natural habitat of captive giant pandas.

Q: What diseases should be paid attention to when giant pandas are released into the wild?

A: The biggest threat to wild population of reintroduced pandas is the spread of diseases to the wild, especially infectious diseases, which may cause devastating disasters to wild populations. Therefore, before the release, it should be determined whether the released individuals are healthy or not, and a comprehensive physical examination must be carried out for this purpose, and all captive giant pandas must be vaccinated against infectious diseases they have ever suffered. In zoos or breeding centers, other kinds of wild animals are often raised at the same time, and various bacteria and viruses they carry may spread across animals. Animals living in zoos have adapted to this environment and have immunity and resistance to these bacteria and viruses. Even if a disease occurs, they can be treated in time. However, animals living in the wild may have never been exposed to these bacteria or viruses, so they are unlikely to have the similar immunity.

Q: What is the goal of the wild reintroduction of giant pandas?

A: The goal of reintroducing giant pandas into the wild is: first, to increase the number of pandas in small populations, improve their genetic diversity, and eliminate their risk of extinction; rebuild the giant panda population in the historical distribution area of giant pandas; second, in helping wild populations to survive for a long time, they can provide example for the protective release of other large mammals, maintain and restore natural biodiversity, promote long-term economic and social development at local and national levels, and promote and improve animal protection awareness of the Chinese people. After giant pandas are released into the wild, continuous scientific testing and research will be carried out, including research on habitat use patterns, activity rhythms, foraging behavior, gut microbiota, hormone changes, and diseases.

Q: Can we use the experience of reintroduction of other species on the reintroduction of giant pandas?

A: At present, a large number of wild release studies have been carried out on the near-source species of giant pandas, such as the American black bear and brown bear, which provide useful experience for the release of captive giant pandas.

Q: What research bases are there currently in China to carry out the wild reintroduction projects of giant pandas?

A: Currently, there are Hetaoping Wild Training Base of China Conservation and Research Center for Giant Panda, Tiantai Mountain Wild Training Base of China Giant Panda Conservation and Research Center, Chengdu Research Base of Giant Panda

Breeding, as well as two wild reintroduction bases, Liziping Giant Panda Reintroduction Base and Daxiangling Giant Panda Ecological Adaptive Reintroduction Base.

Q: Is everything prepared for the reintroduction of giant pandas into the wild?

A: Comprehensive analysis shows that the human has been basically prepared for the wild release of captive giant pandas. Preliminary work for the release of giant pandas has already started. In 2005 and 2006, the giant pandas "Sheng Lin No. 1" and "Xiang Xiang" were released successively. The former were a rescued wild panda. Later, the captive giant panda "Xiang Xiang" died less than a year after being released, but it has explored the way for the wild release of captive giant pandas. It is the price that must be paid for future successful reintroduction. "Tao Tao", who was released on October 11, 2012, was in good health after being recaptured five years later.

Q: How many countries in the world are cooperating in the field of giant panda conservation?

A: Currently, 23 institutions in 19 countries including the United States, Japan, Austria, Thailand, the United Kingdom, France, Belgium, Spain, Finland, Germany, the Netherlands, Denmark, Singapore, Malaysia, Indonesia, South Korea, Australia, Russia, and Qatar are cooperating with China in the field of international cooperative research project on giant pandas. There are 71 giant pandas abroad, fulfilling international community's demand to view China's "national treasures", raising public awareness of protection, enhancing the ability to protect endangered species, promoting international cultural exchanges, and playing a positive role in protecting bio-diversity.

Q: What are the main work done under the global giant panda conservation research cooperation projects?

A: In recent years, China has been actively promoting global action for giant panda protection, and scientific research achievements on giant pandas have been shared globally. On the one hand, China has also been actively carrying out international cooperative researches on giant panda protection. At present, 23 zoos spanning 19 countries have joint China in carrying out cooperative research projects on giant panda conservation. As of October 2022, there are 71 giant pandas participating in international cooperative research projects abroad, and 65 giant panda cubs have survived abroad, of which 37 have returned to China as required by the research projects. On the other hand, China is also actively building a global giant panda protection cooperation and exchange platform. Through regular international and domestic conferences on giant panda conservation, such as the International Conference on Giant Pandas, the Cross-Strait and Hong Kong and Macao Giant Panda Conservation and Education Seminar, and the Annual Conference on Giant Panda Reproduction Technology, China has provided opportunities for the international communities to exchange and discuss new methods, technologies and methods, and achievements for giant panda protection. In addition, China has also actively promoted a domestic scientific research cooperation platform, and established the National Forestry and Grassland Administration Key Laboratory of Rare Animal Conservation Biology in Giant Panda National Park and the Sichuan Provincial Key Laboratory of Endangered Wildlife Conservation Biology.

Q: Following China, which country was the next to successfully breed giant pandas in captivity?

A: It was Japan. In 1979, a giant panda at the Ueno Zoo in Tokyo became the the first panda to give birth successfully in captivity outside China.

Q: Under the form of scientific research cooperation, which countries have successfully bred giant pandas?

A: The most successful example should be the cooperation between the Chengdu Research Base of Giant Panda Breeding and the Wakayama Shirahama Wildlife Park in Japan. From 1994 to 2021, a total of 17 giant pandas have been successfully bred under this project, which is by far the largest number of giant pandas bred abroad. Besides, San Diego Zoo, Washington Zoo, Atlanta Zoo in the United States, Madrid Zoo in Spain, Beauval Zoo in France, Vienna Schönbrunn Zoo in Austria, Ueno Zoo in Japan, and Chiang Mai Zoo in Thailand have all successfully bred giant pandas.

Q: What was the name of the first giant panda that was born abroad and then returned to China?

A: The first giant panda born abroad is called "Hua Mei", who was born at the San Diego Zoo in the United States on August 21, 1999, and returned to Wolong, Sichuan at the age of 4.5. Her parents, "Bai Yun" and "Gao Gao", went to the United States in 1996 to participate in the China-US cooperative research project on giant pandas. "Hua Mei" gave birth three times to three twins after returning to China, that's why she was called "hero mother".

Q: IUCN announced that the endangerment level of giant pandas will be lowered from "endangered" to "vulnerable". Will the protection efforts be reduced?

A: In September 2016, IUCN announced at the 6th World Congress on Conservation of Nature that the endangered level of giant pandas would be downgraded from "endangered" to "vulnerable", which raised great attention from the academic circles, management departments and public media. The "downgrading" of pandas reflects the effects of the efforts made by the Chinese government and the positive results achieved in the protection of giant pandas, and reflects the recognition of China's giant panda protection achievements by the international community. Although the conservation of giant pandas has achieved positive results, isolation of panda populations due to habitat fragmentation is still a major challenge for giant panda conservation. The protection and management capabilities of giant pandas need to be further strengthened and improved, and the threats and endangered status of giant pandas cannot be ignored. The authorities of the Chinese government emphasized that the protection level of giant pandas will not be lowered, and the protection efforts will not be weakened. They are still the flagship species and umbrella species under the protection of endangered species in China, and will continue to unremittingly protect wild animals in accordance with the requirements under China's national first-level protection and the Appendix I of the Convention on International Trade in Endangered Species of Wild Fauna and Flora, to strengthen the protection of giant pandas.

Q: Will giant pandas disappear from the earth?

A: Judging from the current situation, the Chinese government attaches great importance to the protection of giant pandas. It not only values the *in situ* and *ex situ* protection of giant pandas, but also established a giant panda national park to effectively protect the integrity and original elements of giant panda habitats, laying a solid foundation. So giant pandas will be with us human beings, and our descendants will have the opportunity to always visit giant pandas whenever they want.

Q: What is the realistic biological significance of protecting giant pandas?

A: The giant panda is a first-class protected animal in the *List of National Key Protected Wild Animals in China*, and it is an obvious "umbrella species". In a certain sense, protecting giant pandas and their habitats means protecting all wild animals, ecological environments and biodiversity in China's six mountain systems, which include Minshan, Qionglai, Liangshan, Daxiangling, Xiaoxiangling and Qinling.

From the perspective of the scientific research value, social influence and its role in promoting ecological protection, keeping pandas as the top species in wildlife protection and investing adequate resource into the protection work will be helpful to promoting the environmental improvement of giant pandas and their habitats and the integrity of plant species and further expands the richness of biodiversity. In particular, the continuous construction and opening of the giant panda habitat corridor has greatly promoted the genetic exchange between different mountain systems and relatively isolated wild giant panda populations, and has played a decisive role in maintaining population health and its stable growth.

Q: What is the prospective biological significance of protecting giant pandas?

A: Although many achievements have been made in the biological research of giant pandas, as a species, people's understanding of it is still far from enough. How much biological information does the giant panda carry? It is still unknown, such as its evolutionary process, adaptation process, and survival strategies, as well as the microbes in their body that have followed it along the evolution. Finding out the survival information of these giant pandas has great biological significance for the protection of their own species and the protection of other animals. What impact giant pandas will have on human society and human health in the future is also unpredictable and immeasurable. For example, breast milk, which plays a key role in the growth of giant panda cubs, contains very important immune factors and many protein components whose functions are not yet known. Perhaps in the future, some research based on these ingredients will reveal some discovery in human health, life extension or beauty that could surprise the world.

Q: Why do we say that protecting giant pandas means protecting us human beings?

A: The wild giant panda habitats are among the 34 "biodiversity hotspots" in the world. The giant panda is also the apex species in these areas. It is also called the umbrella species. The giant panda habitat is also the habitat of many other animals. If we protect giant pandas, other species in the same area are also protected, and the entire ecosystem in the same area will be protected and restored. A good ecosystem can continuously provide us with public goods such as green mountains and green water, fresh air, and the sound of birds and flowers to meet the most basic well-being of human beings. Therefore, it is often said that to protect giant pandas is to protect ourselves.

"小途"（The ways）
是中国林业出版社旗下文化创意产业品牌，
延续中国林业出版社的专业学术特色和知
识普及能力，整合林草领域专业资源，围
绕"自然文化＋生活美学＋未来科技"，从
事内容创作、内容挖掘、内容衍生品运作。
形成出版、展览、文创、融媒体等优质产
品，系统解读科学知识，讲好中国林草故事，
传播中国生态文化，联手公众建立礼敬自
然、亲近自然的生活方式，展现人与自然
和谐共生的无限可能。

图书在版编目（CIP）数据

大熊猫 201 问 / 张和民等编著 . —— 北京：中国林业出版社 , 2022.12
ISBN 978-7-5219-1928-8

Ⅰ . ①大… Ⅱ . ①张… Ⅲ . ①大熊猫—普及读物 Ⅳ . ① Q959.838-49

中国版本图书馆 CIP 数据核字 (2022) 第 193411 号

编著单位：

中国大熊猫保护研究中心

浙江大学生命科学院

中国野生动物保护协会

四川卧龙国家级自然保护区管理局

Top 201
Questions
About Giant Panda
大熊猫 201 问

项目指导：国家林业和草原局野生动植物保护司

图书策划：李凤波　王佳会

策划编辑：吴卉　杨长峰

责任编辑：吴卉　黄晓飞

宣传营销：张东　王思明　杨小红　蔡波妮

书籍设计：DONOVA

彩色插画：芊袆

黑白插画：郑坤

摄　　影：周孟棋　何胜山　邱宇　李传有　谢浩　刘梅　李伟　罗波　中国大熊猫保护研究中心

特邀编创：小途 The ways

电话：(010) 8314 3552

出版发行：中国林业出版社（100009，北京市西城区刘海胡同 7 号）

印刷：北京富诚彩色印刷有限公司

版次：2022 年 12 月 第 1 版

印次：2022 年 12 月 第 1 次印刷

开本：787mm×1092mm 1/32

印张：8.25

字数：100 千字

定价：68.00 元